GENDER, SPACE AND TIME
Women and Higher Education

Dorothy Moss

LEXINGTON BOOKS

A division of
ROWMAN & LITTLEFIELD PUBLISHERS, INC.
Lanham • Boulder • New York • Toronto • Oxford

LEXINGTON BOOKS

A division of Rowman & Littlefield Publishers, Inc.
A wholly owned subsidiary of The Rowman & Littlefield Publishing Group, Inc.
4501 Forbes Boulevard, Suite 200
Lanham, MD 20706

PO Box 317
Oxford
OX2 9RU, UK

Copyright © 2006 by Lexington Books

All rights reserved. No part of this publication may be reproduced, stored in a retrieval system, or transmitted in any form or by any means, electronic, mechanical, photocopying, recording, or otherwise, without the prior permission of the publisher.

British Library Cataloguing in Publication Information Available

Library of Congress Cataloging-in-Publication Data

Moss, Dorothy, 1952–
  Gender, space and time : women and higher education / Dorothy Moss.
    p. cm.
  Includes bibliographical references and index.
  ISBN-13: 978-0-7391-0997-7 (cloth : alk. paper)
  ISBN-10: 0-7391-0997-9 (cloth : alk. paper)
  ISBN-13: 978-0-7391-1451-3 (pbk. : alk. paper)
  ISBN-10: 0-7391-1451-4 (pbk. : alk. paper)
  1. Women—Education (Higher)—Social aspects. 2. Women—Time management. 3. Time perception. 4. Spatial behavior. I. Title.
LC1567.M67 2006
378.1'982—dc22   2005027303

Printed in the United States of America

∞™ The paper used in this publication meets the minimum requirements of American National Standard for Information Sciences—Permanence of Paper for Printed Library Materials, ANSI/NISO Z39.48-1992.

This book is dedicated with love to my grandchildren Kieran and Rhiannon and my niece, Niamh.

Contents

Acknowledgments		ix
1	Introduction: Lines of Enquiry	1
2	Feminist Theory: A Critical Realist Approach	9
3	Theorising Gender, Space and Time	27
4	The Spatial and Temporal Relations of Women's Everyday Lives	47
5	The Spatial and Temporal Relations of Higher Education	61
6	Research through the Prism of Space and Time	77
7	Spatial and Temporal Practices: Frameworks for Action	99
8	Spatial and Temporal Representations: Guidelines for Action	127
9	The Framework and Guidelines of Higher Education	155
10	Women as Centres of Action	171
11	Conclusion: Women Creating Space and Time	193
Appendices		221
Bibliography		249
Index		263
About the Author		273

Acknowledgments

WITH THANKS TO Gordon Crawford who painstakingly read through and advised on the manuscript, Ingrid Richter who helped design some of the research tools and the original research, Julie Pryke who job shared with me for many years and was a constant support, Sheila Scraton and Rebecca Watson who supervised my doctoral thesis and were always constructive and supportive, Daniel Farrar (my elder son) whose technical assistance was indispensable throughout, Simon Op Den Brouw who worked on the graphics for the final manuscript and Callum Crawford (my younger son) who kept me on my toes throughout. Thanks also to Sonya Farrar and Lucy and Craig Richards-Wilde for their love and support, to Katie Funk and Rebekka Istrail for helpful editorial support and to Kate Robinson for the index.

With special thanks to all the women who participated in and assisted with the research with such warmth, openness and humour.

1
Introduction: Lines of Enquiry

"[I] developed a real thirst for it . . . I found my life through education . . . Very self fulfilled with what I was doing."

Respondent Hazel, chapter 8

"My approach is to read loads and then drag it kicking and screaming together and this probably involves re-inventing it all together."

Researcher Log, chapter 11

Background

THIS BOOK EXPLORES CRITICAL ISSUES related to gender, space and time. The first half explores existing social theory and research in order to develop fruitful research conceptualisations of space and time. The second half applies these conceptualisations in research and analysis of women's daily lives. The research was developed from a collaborative research project,[1] the particular focus of which was women's experiences of creating space and time for higher education in relation to home and household. This initial research formed the pilot study for the work published here. Women respondents from the earlier project had repeatedly mentioned issues related to finding sufficient space and time because of their multiple commitments. I decided to carry out a more in-depth study involving a larger cohort of women. The aim of the research was to explore the struggles and achievements of women students in creating space and time for study considering:

- their experiences and perceptions of the overlapping spheres of home, paid employment and leisure/community in relation to higher education;
- the complexity, diversity and commonalties of experience and perception amongst women students;
- the themes of power, constraint and individual agency.

The research soon became a much wider and deeper study into the sociology of gender, space and time. This introductory chapter is in two parts, considering lines of enquiry, concepts and approach.

Lines of Enquiry

The research is informed by three previous lines of enquiry: feminist research into women and higher education; feminist research which theorises from women's experiences; and social research which gives visibility to the concepts of space and time in research design and analysis.

Feminist Research into Women and Higher Education

A dichotomy has emerged through feminist critiques of women and higher education. On the one hand, higher education is seen to perpetuate gendered inequalities through prescribing certain academic and paid employment routes,[2] for example, social perceptions of curriculum and subject area are gendered.[3] On the other hand, research with women students has shown the ways they identify higher education as liberating them from domesticity,[4] or as a means of subverting racist and sexist expectations of them.[5] Research into women and higher education has tended to focus on the way structures of inequality are embedded within the educational institution and associated institutions (for example, employment) and on the way women say they connect experiences of household and education.[6] However, issues significant to feminist research into higher education might lie beyond both the household and the educational institution. Women interviewed in the earlier research[7] spoke of higher education as it related to the complex and interconnected spheres of home, leisure/community and paid employment and of their involvement in reordering a complex range of responsibilities and activities in order to create space and time to study. The research published here explores women's experience of these overlapping spheres during their period of involvement with higher education. Critical questions are asked concerning the wider relations informing their everyday lives which shape the way space and time for studies is both socially and personally produced.

Theorising from Women's Life Experiences

In order to challenge the processes giving rise to women's subordination, feminists stress the centrality of building theory from women's lived experiences. However it is in the ways that experience is selected and theorised that contention lies. Feminist writers have debated the complexity of exploring differences in women's experience whilst retaining an analysis which maintains the visibility of wider power relations. Dorothy Smith[8] and Sherry Gorelick,[9] for example, argue that women's everyday lives reveal the complexity of the way social relations are lived out, but give less insight into the external processes shaping experience, for example the arrangements of capital or patriarchy at a given time and place. However, through problematising everyday experience, it becomes possible to draw lines of enquiry to the significant factors and relationships of power shaping experience. My choice to explore women's experience of higher education as it relates to their other spheres of experience (household, leisure/community and paid employment) gives visibility to this complexity.[10]

In alerting us to the complexity and diversity of women's experiences, feminists draw attention to the limitations of theoretical perspectives which rely on a single academic discipline,[11] or which draw too simplistic a distinction between the public and the private in women's lives.[12] In order to create space and time women are actively involved in integrating, combining and reordering activities in different and overlapping spheres. A public/private dichotomy tends to give less visibility to this complex process. In addition, feminists have argued strongly for the centrality of women's agency in any research agenda. Feminist research is concerned with challenging social conditions and maximising women's political power.[13]

Social Research: Space and Time

By giving the concepts of space and time priority in the research design and analysis, there is the potential for a research approach which relates personal experience directly to a wider set of social, economic and political relationships. Doreen Massey[14] highlights the way that concepts of space within radical geography have become more dynamic. The spatial is socially produced and the social, in turn, spatially produced. Daphne Spain[15] explores the way in which the spatial segregation of women relates to women's access to reproduce power and privilege. Barbara Adam[16] develops similar themes in her analysis of time, arguing that it is imperative that the social sciences incorporate an understanding of time. Time has a multiplicity of meanings and is experienced and organised in different ways. Time is socialised around benchmarks, which might suit some interests more than others[17]; thus it is a medium of power and an outcome of power relations. Space and time may be conceptualised both quantitatively and qualitatively. There is a particular tension for women whose lives cut across the multiple demands of clock led, body led and household time. John Urry[18] has identified a need for social research which explores issues related to ethnicity and gender drawing on space and time as interdependent concepts.

Key Concepts and Approach

In order to gain insight into the ways that women create space and time to study, it is important to understand that space and time are complex variables, with a range of interpretations that intersect in different ways with women's experience. It becomes necessary to consider the ways in which wider power relations are embedded in women's everyday lives, for example through the ways that space and time are both externally regulated, and regulated by those who exert power in the household. It is also essential to address diversity in women's experience, through examining the way negotiations over space and time are shaped and shape women's social position, as for example, working class women or black women. Of central importance is women's agency in actively transforming and reshaping their activities in order to create space and time. These considerations point to the need for broad, flexible conceptualisations of space and time to underpin research.

The chapter outline is as follows. I review feminist materialist and poststructuralist theories in chapter 2 and identify the work of several writers who combine the strengths of both approaches. This approach has been termed *critical realist*.[19] The key elements of this approach are that it stresses the need for reflexive research that gives attention to knowledge production whilst at the same time maintaining the visibility of central regimes of power and inequality.

In chapter 3 I explore the different conceptualisations of action, space and time developed by a range of authors, in particular Barbara Adam,[20] Karen Davies[21] and Henri Lefebvre.[22] I argue that concepts of space and time give visibility to social difference and individual agency whilst maintaining the visibility of major relationships of power in women's lives. From this review and drawing on conceptualisations developed by the above authors, in particular the work of Barbara Adam and Henri Lefebvre, I further develop the three analytical concepts that underpin the research. These are as follows:

- *Spatial/temporal practices:* this concept gives visibility to the dominant ways in which space and time are organised[23];
- *Spatial/temporal representations:* this concept gives visibility to the inanimate, artefacts, symbols, signs, codes of meaning and knowledge systems[24];
- *Centres of action:* this concept provides the potential for a rich analysis of individual experience in its social, spatial and temporal context. Integral to it is the concept of the *multiple time spans*, for example, the times of the environment, industry, the home and the body, which shape people's lives.[25]

These concepts are used in a variety of ways in the research, including developing the research questions, devising research tools and analysing the data.

In order to test the strength of these concepts and to develop detailed research questions, in chapters 4 and 5 I apply them in a further literature review which considers previous feminist research in the areas of paid work, home, leisure and higher education. It is necessary to consider all these areas, as my primary aim is to explore the multitude of changes that occur in women's everyday lives when they enter higher education. Higher education is a process of change in women's lives, which has effects in all the spheres they are involved in. It has been therefore necessary to develop a detailed and integrated analysis in order to explore the interconnectedness of women experience. From this review I develop critical research questions that give visibility to the spatial/temporal relations within which women try to accommodate higher education. These questions are:

- What are the dominant spatial/temporal practices and representations shaping women's everyday experiences whilst studying?
- What actions (physical, mental and emotional) do women of differing social position take in order to integrate higher education into their lives?
- How have space and time shaped the research?
- What are the implications of the researcher's actions for the findings?

These concepts and questions are subsequently applied in the analysis of empirical data from chapter 7 onwards.

In chapter 6 I discuss my research methodology, considering the epistemology, methodology and methods. Ethical issues and issues related to social division and reflexivity are also discussed. I discuss the ways in which the concepts of space and time inform the methodology, in relation to the knowledge base, the research approach, the research tools and the analysis. I discuss the sample of women who participated in the research. The research sample includes forty-six women from two degree programmes and in their final year of studies. They were based at a community college and were studying for degrees in the social sciences. They reflect a variety of backgrounds and experience in terms of their age, colour, geographical heritage, households, place of birth, accommodation, class background, health, sexuality and dis/ability.

As I am concerned with exploring gendered processes, I draw predominantly on a feminist qualitative methodology. There are limitations to research approaches that attempt to quantify the uses of space and time, because of the complex ways these are experienced and conceptualised. Through the use of a feminist qualitative methodology I hope to have ensured that the research retains certain principles. It is concerned with the complex relationship between processes related to gender, wider relations of power and personal experience. I am explicit about the prior theoretical frame; but responsive to and respectful of participants' voices, frames of meaning and emotions throughout the stages of data collection, analysis and writing up. The research is reflexive, being concerned with the ways in which the spatial and temporal relations I am subject to, including my knowledge base, social position and experiences and emotions, inform the research process at each stage.

Chapters 7 to 10 present the empirical data and the research analysis. In chapters 7 and 8 I directly address the research questions from chapter 4. In chapter 7 I develop a profile of women students' experience in relation to the *spatial/temporal practices* of paid work, home and leisure/community. I develop *snapshots* of women's experiences. In chapter 8 I draw on the concept *spatial/temporal representations* to explore the dominant meanings associated with different spheres of experience at a physical, social and emotional level and areas of *emotional dissonance* for women. I develop *landscapes* of women's experience considering physical, social and emotional features. In chapter 9 I address the research questions from chapter 5 and focus on the college, exploring the practices and representations of higher education shaping women's experiences. Following this, in chapter 10, I address questions from both chapters 4 and 5 and consider the actions women took in order to study, both external and internal to the college. I draw on Barbara Adam's concept *centres of action*[26] to explore the *pathways* women took, the patterns of *spatial connection* they made, the *rhythm* of their daily lives and the *places* they created to study in.

Finally, in chapter 11 I review and evaluate the key research findings. I ask what the strengths and limitations of the research are. For example, has my approach genuinely furthered insight into women's life experiences whilst studying? In making spatial/temporal features visible, do other equally significant factors lose visibility? Does the way in which I use these concepts cast light on differences in experience related to social position? What other approaches to conceptualising space and time might have been used? This chapter also includes a reflexive analysis of the researcher's personal log.

The concepts I have drawn on and developed in this research are unique in the way that they are combined and interrelate. They are drawn from pulling together the work of a range of authors and have proven rich and vibrant in allowing research that pays attention to the diversity of women's experiences whilst also maintaining the visibility of major regimes of power in their lives. There is a growth in feminist research into women's wider life experience and the way this connects to the educational process. Also there is a growth in feminist research that explores concepts of space and time. However, there is not, to my knowledge, research that applies spatial/temporal concepts to the sphere of women's lives and higher education in the ways done here. This is not a study of women's experience of higher education as such, but of the ways in which women create change in their lives.

Notes

1. Dorothy Moss and Ingrid Richter, *Women Students: Creating Space and Time for Study*, paper in Community Studies no. 16, Centre for Research in Applied Community Studies: Bradford College, 2000.
2. Gillian Pascall, *Social Policy: A Feminist Analysis* (London: Tavistock, 1986).
3. Kim Thomas, *Gender and Subject in Higher Education* (Buckingham: The Society for Research into Higher Education and Open University Press, 1990).
4. Gillian Pascall and Roger Cox, "Education and Domesticity," *Gender and Education* 5, no. 1 (1993): 17–35.
5. Heidi S. Mirza, "Black Women in Education. A Collective Movement for Social Change," in *Black British Feminism: A Reader*, edited by Heidi Saffia Mirza (London: Routledge, 1997).
6. Rosalind Edwards, *Mature Women Students: Separating or Connecting Family and Education* (London: Taylor and Francis, 1993).
7. Moss and Richter, *Women Students*.
8. Dorothy Smith, *The Everyday World as Problematic: A Feminist Sociology* (Milton Keynes: Open University Press, 1987).
9. Sherry Gorelick, "Contradictions of Feminist Methodology," in *Race, Class and Gender: Common Bonds, Different Voices*, edited by Esther Ngan-Ling Chow, E. Doris Wilkinson and Maxine B. Zinn (London: Sage, 1996).
10. Derek Layder, *New Strategies in Social Research* (Cambridge: Polity Press, 1993).
11. Romy Borooah, Kathleen Cloud, Subodra Seshadri, T. S. Saraswathi, Jean T. Peterson, and Amita Verma, eds., *Capturing Complexity: An Interdisciplinary Look at Women, Households and Development* (London: Sage, 1994).

12. Patricia Hill Collins, *Black Feminist Thought: Knowledge, Consciousness and the Politics of Empowerment* (London: Unwin Hyman, 1990).

13. Liz Stanley and Sue Wise, *Breaking Out Again: Feminist Ontology and Epistemology* (London: Routledge, 1993).

14. Doreen Massey, *Space, Place and Gender* (Cambridge: Polity Press, 1994).

15. Daphne Spain, *Gendered Spaces* (Chapel Hill: The University of North Carolina Press, 1992).

16. Barbara Adam, *Timewatch: The Social Analysis of Time* (Cambridge: Polity Press, 1995).

17. Adam, *Timewatch*, 61.

18. John Urry, "Sociology of Time and Space," in *The Blackwell Companion to Social Theory*, edited by Bryan S. Turner (Oxford: Blackwell, 1996), 391.

19. Barbara Marshall, *Engendering Modernity: Feminism, Social Theory and Social Change* (Cambridge: Polity Press, 1994); Andrew Sayer, *Method in Social Science: A Realist Approach* (London: Routledge, 1992).

20. Barbara Adam, *Time and Social Theory* (Cambridge: Polity, 1990); Adam, *Timewatch*; Barbara Adam, *Timescapes of Modernity: The Environment and Invisible Hazards* (London: Routledge, 1998).

21. Karen Davies, *Women and Time: The Weaving of the Strands of Everyday Life* (Aldershot: Avebury, 1990).

22. Henri Lefebvre, *The Production of Space* (Oxford: Blackwell, 1991).

23. Developed from the work of Lefebvre, *Production, Space* in relation to the concept *"spatial practices."*

24. Developed from the work of Lefebvre, *Production, Space* in relation to the concept *"spatial representations."*

25. Adam, *Time, Theory*, 66.

26. Adam, *Time, Theory*, 66.

2

Feminist Theory: A Critical Realist Approach

"When I was working on the farm I struggled with my femininity . . . I wanted to be petite and dress nicely. So when I did go out I was always wearing dresses and showing off my attributes (both laugh) and then I got, I was giving people the wrong messages you know. Oh terrible you know, to be feminine. I wanted to be one of these girls; I was always impressed by these girls who always had a handbag. They had tissues in their handbag."

Respondent Rian, chapter 8

"I'm grappling with feminist theory paper. Keep getting an exciting framework and then not sure I grasp the ideas—particular meanings of specific terms, e.g. structuralist/post-structuralist. Think I've got it then it disappears like smoke."

Researcher Log, chapter 11

Introduction

IN THIS CHAPTER I IDENTIFY the theoretical debates which underpin my research. I point to the need for conceptualisations of social and personal change which are broad enough to give insight into the wider social relations shaping women's lives yet retain the flexibility to give visibility to differences between women and their personal actions. The chapter is in three parts considering: materialist analyses of women's oppression; post-structural analyses and feminism; and feminist critical realist approaches. Initially two questions arise. What is meant by feminist theory? What characterises the relationship between feminist theory and feminist research?

Feminist theorists attempt to develop concepts to explore, explain and challenge women's oppression. Theory is clearly not neutral and objective: it is shaped by and shapes its own milieu. Whereas Shulamit Reinharz[1] sees feminist theory as guiding feminist research, Liz Stanley[2] would not privilege theory in this way. She positions academic knowledge production squarely within the academic market, and argues that it involves all the distortions of the production process inherent in that. Theory is shaped by specific academic disciplines, the sexual, social and academic divisions of labour, and the beliefs, values and experiences of the theorist. Stanley argues that knowledge production is one site of feminist struggle and that feminist theory does not necessarily guide feminist research. Research practice may well challenge theoretical paradigms.

Janice Moulton[3] argues that developments in theory have been shaped by the *adversary method* in philosophical reasoning, that is, that the best method of testing claims of truth is to challenge them. She argues that this method involves moving onto the ground of the adversary. In so doing, the paradigms of the adversary are inevitably adopted and the parameters of debate become narrowed. A non-adversarial approach would facilitate the development of paradigms more directly suited to the purpose in hand. The *adversary method* is primary in scientific reasoning and has been highly influential in shaping philosophical debate. The development of feminist theory is also characterised by the *adversary method*. For example, the development of a materialist analysis of women's oppression relies heavily on existing paradigms from structuralist theory and particularly from Marxism. The debate between post-structuralists and feminists involves reworking paradigms developed by French post-structuralist writers. However, a further characteristic of feminist theory is the way writers challenge previous ideas by enlisting knowledge from other areas. For example, they tend to integrate ideas from a number of academic disciplines concerned with both macro and micro spheres of knowledge. They also give attention to knowledge and ideas from outside the academic sphere including feminist activists and women with particular experience. For example, Patricia Hill Collins[4] draws on the insights of a range of black women to develop *an afro-centric feminist epistemology* including the knowledge of both academics and non-academics.

Materialist Analyses of Women's Oppression

By *materialist analyses* I am referring to writers who give central significance to the *material* in shaping social relations. This may include production, reproduction, sexual divisions and the body.[5] Second wave feminist writers drew on Marxist theory and method to develop paradigms that could facilitate exploration of women's position. The Marxist focus on the economic and the division of labour was influential in Shulamith Firestone's "The Dialectic of Sex,"[6] in which she argues that it is in fact the sexual division (based in biology) that forms the root of other social and economic divisions including the division of labour. Women's position in relation to birth, reproduction and the nurture of infants should be considered *"fundamental—if not immutable—fact."*[7] *"I have attempted to take class analysis one step further."*[8] Thus, Firestone draws on Marxist historical method to theorise sexual divisions. Because of her emphasis on women's biological position, she has been criticised as biologically determinist. However, it is clear that at the time of writing (1970) she did not see these biological divisions, when represented in their social form, as inevitable. She held out a utopian alternative for women. By extending Marxist analytical method to consider sexual divisions, she raised basic questions about the adequacy of the Marxist view that class divisions are central in shaping social relations.

Gayle Rubin[9] also addresses inadequacies in Marxist theory in relation to the question of women. She provides a feminist reading of Marx and draws on two other theoretical schools. By bringing different paradigms together she furthers an explanation of women's material exploitation. She argues that Marxist positions in relation to women and the division of labour do not fully explain why women are socially positioned as they are. An analysis that views women as simply a reserve army of labour for capitalists is deemed insufficient as are analyses that emphasise women's management of capitalist consumption for the household.[10] Rubin argues that housework should be conceived of as *"a key element in the reproduction of labour power."*[11] Housework sustains and reproduces the labour force and is directly connected to the wider economic sphere. Whether housework should be conceived of as *production* or *reproduction* within economic analysis has remained a debated issue for feminists.[12] Rubin argues that the question as to *why* women are socially positioned in particular ways (for example, in the domestic sphere), is not satisfactorily explained through Marxist economic theory. She draws on anthropological, kinship and psychoanalytic theory to provide a more complete explanation. Drawing on Lévi-Strauss she discusses the traffic in women within kinship systems[13] and drawing on Lacan she explores the ways in which biological sex has cultural meanings which are internalised by individuals and culturally reinforced.[14] She develops the idea of *"sex-gender systems"* that are *"products of historical human activity,"*[15] and that connect women's position within the spheres of household, community and work. Through these complex systems women's oppression is sustained.

By drawing together a range of paradigms, Gayle Rubin's work is characteristic of many feminist writers who have been concerned to reinterpret and challenge existing bodies of knowledge. However, in this case, the economic, anthropological and psychoanalytic paradigms she draws on sit awkwardly together. The *adversary method* informs her work as she tackles each field of knowledge, and reinterprets it from a feminist perspective. In each case she adopts existing paradigms in order to challenge them and the problems associated with those paradigms sometimes remain in her work. For example, the separation of the concepts of the *economic* from the *social* in Marxist theory[16] (see Wendy Bottero, below) and the emphasis in psychoanalytic theory on the parent-child bond and unconscious processes as having primary significance in shaping social relations have been challenged as questionable ideas from a feminist perspective.[17]

It has been argued that the concept of patriarchy aids analysis of women's oppression: *"a set of social relations between men, which have a material base, and which though hierarchical, establish or create interdependence and solidarity among men that enable them to dominate women."*[18] This definition of patriarchy echoes Marxist understanding of the relations of the capitalist over the working class. The concept both challenges the adequacy of Marxist concepts and draws on Marxist understanding of historical process, material position, structures of power and ideology. Heidi Hartman argues that patriarchy and capitalism are distinct hierarchical systems, which may work in partnership, through, for example, the establishment of the family wage.[19] Achieving an adequate wage for the head of household to support the whole family is central to economic bargaining. This has served both the interests of men and of capitalists, having diminished women's economic bargaining power. Hartman also refers to a racial hierarchy[20] and thus highlights three distinct structures of inequality based on class, gender and "race," although she does not discuss the latter in detail.

Nancy Hartsock[21] draws on Marxist theory in relation to concepts of knowledge and ideology. Again, she both challenges and reinterprets existing theory from a feminist perspective. She argues that because the position of women is *structurally different* to that of men, there are inevitably epistemologological consequences. Capitalists are concerned with the *exchange value* of the product of labour rather than its *use value*. This involves them in distorting and inverting reality. *Use value* is not their central concern, rather the *surplus value* which can be extracted for profit accumulation. Here are the origins of bourgeois ideology.[22] Marx had argued that the proletariat was able to *see under* bourgeois ideology because of their specific relationship to the production process. Hartsock reinterprets these ideas from a feminist perspective and argues that women are involved in both production and reproduction and are able to "*see under... [an] ... abstract masculinist ideology*"[23] developed in the interests of men who are concerned to extract benefit from women's activity and bodies.[24] Men's relationship to production is distorted because of the sexual division and their separation from production in the home. This ideology distorts reality in the interests of men and the hierarchical dualities which shape this ideology, for example, *mind/body, idea/material, social/natural, reason/emotion,* serve to diminish women and sustain their exploitation. Women are inevitably associated with the inferior side of the duality. A *feminist standpoint* or epistemology may see under these ideological masculinist constructions.[25] The assumption here is that a feminist standpoint is able to give a different and more scientific account of the real world: *"An analysis which begins from the sexual division of labour... the real material activity of concrete human beings ... could form the basis for an analysis of the real structures of women's oppression."*[26]

Feminists have debated the adequacy of the concept of patriarchy as a hierarchical system of inequality based on sexual divisions. It has been argued that the concept lacks both historical[27] and geographical[28] specificity. Michele Barrett,[29] for example, argues, "*Masculinity and femininity obviously are categories of meaning in one sense, but men and women occupy positions in the division of labour and class structure which although not pre-given, are historically concrete and identifiable.*"

The concepts of patriarchy and capitalism describe separate (albeit overlapping) systems of inequality. There have been heated debates between feminists as to their relative significance for the oppression of women. Hartmann's[30] concept of *dual systems* is a means of giving equal weight to capitalism and patriarchy in understanding social relations. This concept has however been contested. By posing these as distinct spheres of oppression, Iris Young asks, does not patriarchy become a biological backdrop to capitalism? This is because the primary focus of the concept is on family and kinship which results in a trivialisation and simplification of women's oppression: "*It seems reasonable ... to admit that if capitalism and patriarchy are manifest in identical social and economic structures they belong to one system, not two.*"[31] The focus of dual systems is on two static structures rather than process, but in reality "*what we are seeing is struggle that contributes to the formation of people within both their class and gender simultaneously.*"[32] Cynthia Cockburn argues for a focus on the *material* and *physical power* as opposed solely to the economic. These concepts bridge the systems of oppression and give visibility to the range of human experience: "*I use the term physical power to mean both corporal effectivity (relative bodily strength and capability) and technical effectivity (relative familiarity with and control over machines and tools.)*"[33] The search for concepts that bridge the social divisions has been further advanced through black feminist and anti-racist analysis.[34] There are great difficulties involved in making connections between hierarchical systems theorised separately around the concepts of class, "race" and gender. Rose Brewer[35] argues that the concept of *simultaneity* takes forward understanding of the ways formations of class, "race" and gender interrelate in shaping the position of black women in the labour market. Questions arise as to the adequacy of the focus on *structure*[36] in Marxist and feminist theory to account for the complexity of relationships of power. Barbara Marshall argues that the idea of a single feminist standpoint is insufficient to account for different axes of power related to a complex range of economic and social divisions.[37] Many forms of reproduction *"take place outside the social relations of production theorised by Marx."*[38] The concepts of *multiple standpoints*[39] and *multiple* and *shifting subjectivities*[40] have been proposed to resolve these issues.

Materialist analyses of women's oppression provide a wide range of analytic tools to address the way that the wider material environment, including the organisation of paid and unpaid work, the home and leisure relate to divisions of class, "race" and gender. It is important to sustain these insights in examining the daily lives of women. However, the difficulties in explaining the relationships between structures of power, systems of knowledge and individual agency led feminists to turn to theory which might bridge the concepts of structural power and individual action. For example, Beverley Skeggs[41] draws on Bourdieu's concepts of economic, cultural, social and symbolic capital (as *traded* in social interaction) to give an account of social formations relating to class and gender and Dorothy Smith[42] argues that problematising the *everyday world* of women allows entry into the totality of complex material relations, both external and internal, which shape their lives. The limitations of structuralist theory to account for the complex positioning of women, for the relationship between structures of power and individual agency and for the relationship between knowledge and ideology led some feminists to look to paradigms from post-structuralist theory.

Post-Structural Analyses and Feminism

Nancy Fraser[43] clarifies the threads of theory informing the post-structural school. She argues that two strands are evident. One draws on Foucault's historical method and the other is associated with the work of Derrida and Lacan. Foucault developed concepts to enable exploration of the way social power operates. Central to his work are theories of *discourse* and *claims to truth*.[44] To explain the exercise of power it is necessary to challenge two ideas. Firstly the idea that one social group *holds* power over another and secondly the idea that ideology and knowledge are separate.[45] Power penetrates social organisation in complex ways and through different levels of events: *"There are actually a whole order of levels of different types of events differing in amplitude, chronological breadth, and the capacity to produce effects."*[46] Thus, the social is regulated through the construction of *claims to truth*.[47] A *discourse* describes the boundaries whereby things in the world are socially understood and involves language as well as social and institutional arrangements. In *Madness and Civilisation*,[48] Foucault argues that the discourses around *lunacy* (including institutional arrangements and medical interventions) are essentially concerned with describing and defining what should be considered as sane and rational in society. In his work on punishment, he argues that discourses about punishment did not emerge to repress wrongs, but to define wrongs. Systems of punishment arise from concrete social, political and moral conditions.[49] Discourses about sexuality in the nineteenth century were not concerned to limit certain types of sexual behaviour but served to define a socially acceptable sexuality.[50] In order to understand the operation of power in society, one should consider how a particular discourse arises, where lines of influence lie, what is included and what is excluded from the discourse. Foucault's analytical method is historical and *archaeological*, that is it involves unpicking the accumulation of practices and *"the genealogy of relations of force,"*[51] which give rise to specific discourses.

For feminists the attractions of Foucault's approach are evident as it enables consideration of the diversity of women's experience, the impact of social position and the way discourses related to women are constructed. Although Foucault considers the ways that sexuality is socially constructed, he does not address the way gender divisions inform that process. Fraser, however, argues that his method does enable us to see *"how identities are fashioned over time, how under conditions of inequality groups are formed and unformed."*[52] Foucault breaks with the concept of unitary structures of power and the distinction between ideology and knowledge in Marxist theory. Fraser argues that Foucault's approach is very different to the approach within post-structural theory associated with Derrida and Lacan (linguistic and psychoanalytic respectively). Derrida's position is that it is not possible to access social reality. All we have access to is language as an account of that reality and as a medium through which we interpret reality. The object of our studies should be *the text*.[53] Lacan focuses on identity formation, the centrality of symbolic structures and the role of the unconscious.[54] Fraser[55] argues there are limitations to post-structural approaches that overemphasise language and/or base themselves on *"autonomous psychological imperatives"* at the expense of historical specificity. She argues that historically specific social practices (discourse) should be the focus of analysis.

Supporting a post-structural approach, Samantha Ashenden[56] argues that it is attractive because of the inadequacies of structuralist theory and because of the insight gained into the way that power operates at the local level. This benefits feminism, as we are able to unearth the specific ways in which discourses about women arise, and in particular the ways that gender is constructed and contested. There are several examples where analysis drawing on post-structural approaches appears to strengthen feminist politics. By breaking with the ideology/knowledge dichotomy, Judith Butler[57] explores the complexity involved in the process of *coming out*. People may choose to come out in resistance to homophobia, but the identity *lesbian/gay* actually emerges from a homophobic discourse. By claiming these identities as positive affirmations, she asks whether we perpetuate regulatory regimes or truly liberate ourselves. She then explores the complex ways in which people imitate, and parody hegemonic (dominant) notions of sexuality.

In a similar way, Paul Connolly[58] draws on post-structuralist ideas to explore questions of racialisation. He argues that post-structuralism facilitates research grounded in experience. Rather than focusing on *black* and *white* subjects, it is more strategically useful to explore the way that things in the world are racialised in particular ways at particular times. This has the potential to strengthen anti-racist strategy. For example, anti-racist feminists have identified the complex ways in which discourses relating to colour have been applied in social analysis. Ruth Frankenburg[59] argues that the category of *whiteness* has been insufficiently problematised. Where black women's experiences have been discussed, the concept *black* has been problematised. *Whiteness* is either assumed to be the norm or remains invisible. As a result, research has either excluded black women's experience as at all different, for example, research into women's unpaid caring activities has downplayed the different stratification of black women's employment and their historic role in paid domestic work.[60] bell hooks[61] writes of white feminists' *"deep emotional investment in the myth of sameness,"* and of *"white control of the black gaze."* Such analysis exposes the complex ways in which white superiority is socially sustained.

The key argument from post-structuralism is that social structures are constituted through complex relationships and events.[62] The social should be viewed as relational and not structural. The strength of post-structuralist approaches is that they invite deep questions of the categories of knowledge being employed and facilitate creative ideas and different political strategy. However its weakness for feminists rests in the denial of major structures of power, of a *grand narrative* or of *"the search for the fundamental causes of injustice."*[63]

Feminist Critical Realism

The concept of feminist critical realism draws on Andrew Sayer's concept of realism: *"Knowledge and practice are interdependent and knowledge is embedded in social practice. . . . Theory does not order given observations or data but negotiates their conceptualisation."* Research should not be "fact gathering" but should be concerned with *"the reduction of illusion and emancipation."*[64] Barbara Marshall[65] employs a similar concept, that of critical modernism, which, she argues, is *"non-positivist, critical of the hegemony of western research, considers local stories, reconstitutes 'subject,' rejects universality or grand narratives."*

The debate within feminist theory between post-structuralists and structuralists/materialists (including feminists who identify with both radical and socialist traditions) reached an impasse in the 1990s. It had been alleged that structuralist theories had failed to account adequately for women's oppression and that the strength of post-structuralist concepts was that they give visibility to and contested the ways that the knowledge of *woman* is constructed. Although a concept common to all feminist approaches and not specific to post-structuralism, the concept of *gender* is particularly elevated in post-structuralist approaches, as it shifts the focus from women's experience to the gendered relations giving rise to experience: *"The dissolution of the category 'woman' as an epistemological referent is not the same as arguing that gender is not a category of oppression."*[66]

However, the conflict at the Beijing conference (1995) conveyed the seriousness of the impasse between feminists.[67] Conservatives from the North attacked the concept of *gender* arguing that it facilitated a deviant attack on women who were naturally oriented to family. In addition, some feminist activists from the South also attacked the concept of *gender* but argued it actually facilitated the de-politicisation of issues because the field of gender analysis was both overly complex and silenced activists from the South. Women's experiences of oppression were no longer the direct focus and the gender focus was being used to *"deny the very existence of women specific disadvantage."*[68] Although some feminists believed that the field of gender studies had opened the door to more radical possibilities for feminism, it was argued that its appropriation by other forces involved in international policy making appeared to lead to diversions from a feminist and woman-centred agenda. The term *phallic drift* has been coined to convey the ways in which highly abstract theories may allow the male perspective to re-assume precedence: *"the powerful tendency for public discussion of gender issues to drift inexorably back to the male point of view."*[69]

At the level of theory, there has been an impasse between those who believe that the world is knowable, predictable, and changeable through reflexive, scientific research and those who stress the relativity of knowledge. Pierre Bourdieu[70] and Sandra Harding[71] for example, argue that it is essential to explore the specific relations giving rise to social facts, but that this enterprise *is* scientific. Michel Foucault[72] argues that it is not possible to stand outside our own knowledge system; that all knowledge is relative and therefore it is not possible to apply scientific method to change the world.

Recently, several feminist writers have tried to bridge this impasse at the level of both theory and political action. There have been attempts to draw on the strengths of both post-structuralist and materialist schools of thought in order to move the feminist agenda forward. This tendency has been evident in the development of certain themes:

- real world research is possible;
- the identification of central regimes of power and resisting fragmentation are important;
- approaches which are reflexive and give attention to knowledge production are central;
- new theoretical and political concepts that give visibility to complex power relations and further a broad feminist agenda are needed.

Real World Research Is Possible

Sue Clegg[73] suggests that it is impossible to resolve the philosophical impasse concerning the accessibility of the real world to research. It is possible however to adopt a realist *ontology* (or *feminist objectivity*), a belief that we can conceptualise the real and that there are mechanisms in nature with real powers which we cannot observe but which are open to scientific enquiry. *Critical realism* allows for a reflexive yet scientific approach. Linda Alcoff discusses the impasse between those who demand a research focus on women's real historical experiences, and those who argue that *woman* is a fiction and the way this fiction is created should be the subject of research. Alcoff argues, *"One's subjectivity becomes engendered through constant interaction with the world . . . discursive boundaries of consciousness change with historical conditions."*[74] She makes the case for *positionality* as an approach in feminist analysis and argues that political theory should be based on *"the initial premise that all persons, including the theorist, have a fleshy material identity that will influence and pass judgement on all political claims."*[75] An awareness of position avoids the essentialism inherent in the idea that *woman* has some fixed identity and culture. The concept highlights political movement and agency.[76] Both the ideas of woman as fictional construct and women as real historical beings have relevance,[77] as it is the intersection between these that is the focus of study. Both

Clegg and Alcoff discuss concepts that give visibility to real historical experience without abandoning some of the critical issues raised by post-structuralists.

The Identification of Central Regimes of Power and Resisting Fragmentation Are Important

"How do we negotiate the treacherous course of rejecting the fiction of 'woman' as a fixed category, while at the same time recognising the need to fight for the particular rights of women?"[78] Feminist writers have debated the complexity of exploring differences in women's experience whilst retaining an analysis which maintains the visibility of structures of power. Sherry Gorelick,[79] for example, argues that women's experiences reveal the complexity of the way social relations are lived out, but give less insight into the external processes shaping experience, for example the arrangements of capital or patriarchy at a given time and place. The danger of locating central significance on local understanding has led to overemphasis on experience and identity at the expense of the forces shaping these. Nira Yuval Davis[80] argues that the retreat from exposing major structures of power or *grand narratives* is very dubious, particularly at a time when neo-conservative and fundamentalist forces are gathering strength. These writers draw feminists' attention to the need to identify wider regimes and systems that shape women's experience.

Approaches That Are Reflexive and Give Attention to Knowledge Production Are Central

Writers like Liz Stanley[81] refuse to identify themselves with any particular school of thought but stress the centrality of reflexivity in the research process; that is the importance of being critically aware of the conditions in which feminist knowledge is produced and the need to situate self in the research process. They also stress the importance of including emotion as part of the research agenda. Emotion has been discredited as a route to insight, being positioned as inferior to *reason* in the dualisms of "classical" philosophy. Nira Yuval Davis[82] uses post-structuralist ideas to question the way that fundamental categories of knowledge (*nation, state, family, culture, citizenship*) have been constructed in gendered and interconnected ways. Echoing Marshall's concepts of the *subject in process*,[83] Yuval Davis argues that the strength of post-structuralism is that it draws attention to the fact that the subject is social[84] and that our perceptions of the real world are shaped by our knowledge base. For example, in relation to culture she argues, *"Women are often constructed as the cultural symbols of the collectivity, of its boundaries, as carriers of the collectivity's 'honour' and as its intergenerational producers."*[85]

Thus, Stanley, Wise, Marshall and Yuval Davis all draw on strengths from post-structuralism. They stress the relativism of knowledge and the ways in which feminist strategy may be strengthened by questioning categories of knowledge. However, they do not abandon the search for the (albeit fractured) foundations of women's oppression. Examining and unpicking the complex power relations producing knowledge of women is seen as a necessary first stage to reconstructing an effective critique of gender divisions. As Barbara Marshall argues,[86] *"Attention to the discursive construction of subjectivity and the plurality of subject positions need not mean that a material level of analysis is eschewed. This requires coupling deconstruction with reconstruction."* It is through the process of reconstruction that central regimes of power become visible in different women's lives.

New Theoretical and Political Concepts That Give Visibility to Complex Power Relations and Further a Broad Feminist Agenda Are Needed

The concepts of *patriarchy, capitalism* and *racial hierarchy* describe separate but overlapping systems of oppression. The contradictions and omissions arising from these conceptualisations were discussed above ("Materialist Analyses of Women's Oppression"). Nevertheless, there is no doubt that the development of these concepts and the historical method associated with them, furthered insight into formations of gender, class and "race." Post-structuralist concepts such as *discourse, text, claims to truth* and the *archaeology of knowledge* (discussed in "Post-Structural Analyses and Feminism") may strengthen political understanding of the operation of power at a local and situated level, but provide no overall solution to the search for the foundations of inequality. Feminists have been developing concepts that address this impasse at both the theoretical and political level. They have been seeking concepts that will facilitate understanding of the relationship between the local and global, structure and agency, the researcher and the researched and yet remain centrally concerned with the search for social justice. For example, Fiona Williams[87] has developed the concept of the *polyhedron* to represent diversity and power. This is built on the axes of class, age, disability, gender, "race" and sexuality. It is informed by the relations of work, nation and family, and by the divisions of patriarchy, capitalism and imperialism. The *polyhedron* draws on both structuralist and post-structuralist ideas. It represents major structures of power but facilitates understanding of complex axes of power and position. The intention is to enrich social policy analysis through its use. Wendy Bottero accepts the major criticisms of structural theory raised by post-structuralists, but argues that this awareness does not necessarily lead us to abandon the search for structural explanations or to the fragmentation inherent in post-modern theory. Bottero argues that a fundamental flaw in Marxist theory is the separation of the idea of the *economic* from the *social* and the development of distinct concepts to explain this. She argues that *patriarchy* and *"race"* have been theorised as systems which position women and black people socially. Capitalism is seen as a morally neutral profit driven system solely associated with the *economic*. She argues that the idea of capitalism and the market as neutral, value free, purely for profit processes is the source of some of the difficult debates with feminism. Feminists have tended to accept this perspective on the economic and as a result have had difficulty developing a clear account of women's exploitation. Bottero argues that values, claims and obligations relating to class, "race" and gender inform economic processes and choices. Having argued that previous conceptualisations of the economic and social as separate spheres are inadequate, she suggests that, *"Instead of seeing multiple and overlapping structures of inequality . . . we can theorise a single, structural social space. This space—composed of social interactions—can be seen as a structure of social distance, similarity and dissimilarity. . . . Individuals located at different points in that space . . . will be engaged in a great variety of social relations."*[88] By focusing on the concept of *social space*, both Williams and Bottero try to address some of the theoretical impasse between structuralist and post-structuralist schools. They develop more holistic concepts to facilitate understanding of the complexity of power relations.

There has also been an impasse in relation to feminist political strategy. Materialist feminists stress the importance of *standpoint*. Together with the idea that the *personal is political*, this gives credence to the idea that true insight can only be developed from personal experience and insight will vary according to social position. Some women's accounts of reality may be considered more valid than those of other women because of their social position. From a post-structuralist perspective *subjugated discourses* are considered important sites for research. These are discourses that are excluded but inform the shape of the dominant discourses. Credence is thus given to the voices of the marginalised and oppressed in both post-structuralist and materialist perspectives. The danger of slipping into an identity-based politics is evident in both schools of thought. Yuval Davis has argued that identity politics is unproductive as identities are socially constructed[89] and constructed across social difference.[90] She argues that *dialogue* is the basis of empowered knowledge and discusses *transversal politics* as a progressive position. The concept of *transversalism* is developed from the ideas of the autonomous left in Bologna. It provides an alternative to either universalist or relativist positions. Universalism may lead to exclusionary practices, for example ethnocentrism. On the other hand relativism may involve an identity-based politics and lead to fragmentation and sectarianism. Yuval Davis, citing Assiter, argues that a transversal politics "*differentiates between social identities and social values and assumes that . . . 'epistemological communities' . . . which share common value systems can exist across differential positioning and identities.*"[91] She argues that individuals, whether writers, researchers or activists, do not represent particular groups, but may advocate on their behalf.

Conclusion

The authors discussed in the last section ("Feminist Critical Realism") move beyond both the theoretical and political impasse in feminism discussed previously. They draw on structuralist and post-structuralist ideas, which facilitate understanding of power, gender and political action. Having reviewed these strands within feminist thought I have developed my research from a feminist critical realist perspective reflected in the following elements. Attention is given to the processes giving rise to the research. Different schools of thought are drawn on, as appropriate and the connections between them identified, for instance, from both materialist and post-structural perspectives. Visibility is given to the differing social position of women including their age, household status, class, colour,[92] geographical heritage and other ascribed and self-chosen identities. The significance of social position is explored. The research problematises everyday experience[93] and explores women's actions in all the spheres they occupy including paid work, home and leisure. Links are made to the "*behind scenes of power and control*" shaping action.[94] Research tools and concepts are drawn on which facilitate all the above, in particular the concepts of space and time. Wendy Bottero and Fiona Williams draw on spatial concepts in developing their ideas, including the concepts of *social space* and the *polyhedron*; however the meanings attached to such concepts are complex. In the next chapter I move on to a discussion of the concepts of space and time. I will argue that by making these more visible in feminist research some of the political and theoretical difficulties discussed in this chapter may be resolved.

Notes

1. Shulamit Reinharz with Lynn Davidman, *Feminist Methods in Social Research* (New York: Oxford University Press, 1992), 240.
2. Liz Stanley, "Feminist Praxis and the Academic Mode of Production," in *Feminist Praxis*, edited by Liz Stanley. (London: Routledge, 1990).
3. Janice Moulton, "A Paradigm of Philosophy: The Adversary Method," in *Women, Knowledge and Reality, Explorations in Feminist Philosophy*, edited by Ann Garry and Marilyn Pearsall (London: Routledge, 1996).

4. Patricia Hill Collins, *Black Feminist Thought: Knowledge, Consciousness and the Politics of Empowerment* (London: Unwin Hyman, 1990), 206–20.

5. Sonya Andermahr, Terry Lovell and Carol Wolkowitz, *A Concise Glossary of Feminist Theory* (London: Arnold, 1997), 133.

6. Shulamith Firestone, "The Dialectic of Sex," in *The Second Wave: A Reader in Feminist Theory*, edited by Linda Nicholson (London: Routledge, 1975/1997).

7. Firestone, "Dialectic, Sex," 23.

8. Firestone, "Dialectic, Sex," 25.

9. Gayle Rubin, "The Traffic in Women: Notes on the 'Political Economy' of Sex," in *The Second Wave: A Reader in Feminist Theory*, edited by Linda Nicholson (London: Routledge, 1975/1997).

10. Rubin, "Traffic, Women," 28.

11. Rubin, "Traffic, Women," 30.

12. Romy Borooah, Kathleen Cloud, Subodra Seshadri, T. S. Saraswathi, Jean T. Peterson and Amita Verma, eds., *Capturing Complexity: An Interdisciplinary Look at Women, Households and Development* (London: Sage, 1994), 50.

13. Rubin, "Traffic, Women," 35–42.

14. Rubin, "Traffic, Women," 43–55.

15. Rubin, "Traffic, Women," 55.

16. Wendy Bottero, "Clinging to the Wreckage? Gender and the Legacy of Class," *Sociology* 32, no. 3 (August 1998): 469–90.

17. Andermahr, Lovell and Wolkowitz *Glossary, Feminist*, 67 and 54.

18. Heidi Hartmann, "The Unhappy Marriage of Marxism and Feminism: Towards a More Progressive Union," in *The Second Wave: A Reader in Feminist Theory*, edited by Linda Nicholson (London: Routledge, 1975/1997), 101.

19. Hartmann, "Marxism, Feminism," 105.

20. Hartmann, "Marxism, Feminism," 101.

21. Nancy Hartsock, "The Feminist Standpoint: Developing the Ground for a Specifically Feminist Historical Materialism," in *The Second Wave: A Reader in Feminist Theory*, edited by Linda Nicholson (London: Routledge, 1983/1997).

22. Hartsock, "Feminist Standpoint," 220.

23. Hartsock, "Feminist Standpoint," 227.

24. Hartsock, "Feminist Standpoint," 222.

25. Hartsock, "Feminist Standpoint," 220; Sandra Harding, *The Science Question in Feminism* (Milton Keynes: Open University Press, 1996).

26. Hartsock, "Feminist Standpoint," 233.

27. Sue Clegg, "The Feminist Challenge to Socialist History," *Women's History Review* 6, no. 2 (1997), 201–14.

28. Nira Yuval Davis, *Gender and Nation* (London: Sage, 1997), 7, citing Moghadarm.

29. Michele Barrett, "Capitalism and Women's Liberation," in *The Second Wave: A Reader in Feminist Theory*, edited by Linda Nicholson (London: Routledge, 1997), 126.

30. Hartmann, "Marxism, Feminism."

31. Iris Young, "Beyond the Unhappy Marriage: A Critique of Dual Systems Theory," in *Women and Revolution: The Unhappy Marriage of Marxism and Feminism—A Debate on Class and Patriarchy*, edited by Lydia Sargent (London: Pluto Press, 1981), 47.

32. Cynthia Cockburn, "The Material of Male Power," in *Waged Work: A Reader*, edited by Feminist Review Collective (London: Virago, 1986), 94.

33. Cockburn, "Male Power," 97.

34. bell hooks, *Feminist Theory: From Margins to Centre* (Boston, MA: South End Press, 1984); Hill Collins, *Black Feminist*; Stanlie M. James and Abena P. A. Busia, eds., *Theorising Black Feminisms: The Visionary Pragmatism of Black Women* (New York: Routledge, 1993).

35. Rose Brewer, "Theorising Race, Class and Gender: The New Scholarship of Black Feminist Intellectuals and Black Women's Labour," in *Theorising Black Feminisms: The Visionary Pragmatism of Black Women*, edited by Stanlie M. James and Abena P. A. Busia (London: Routledge, 1993), 16.

36. Roy Boyne, "Structuralism," in *The Blackwell Companion to Social Theory*, edited by Bryan S. Turner (Oxford: Blackwell, 1996). He argues that s*tructure* refers to the " . . . *theorisation of larger forces* . . . " which are external to individuals and which shape social forms and delimit action.

37. Barbara Marshall, *Engendering Modernity. Feminism, Social Theory and Social Change* (Cambridge: Polity, 1994), 27 and 29. She argues that it is simplistic to reduce explanations of social relations to labour-capital relations. She makes the case for a perspective that focuses on the relationship between the individual and society.

38. Terry Lovell, "Feminist Social Theory," in *The Blackwell Companion to Social Theory*, edited by Bryan Turner (Oxford: Blackwell, 1996), 311.

39. Liz Stanley and Sue Wise, *Breaking Out Again. Feminist Ontology and Epistemology* (London: Routledge, 1993).

40. Norma Alarcon, "The Theoretical Subject(s) of This Bridge Called My Back and Anglo-American Feminism," in *The Second Wave: A Reader in Feminist Theory*, edited by Linda Nicholson (London: Routledge, 1997); Elsa B. Brown, "What Has Happened Here? The Politics of Difference in Women's History and Feminist Politics," in *The Second Wave: A Reader in Feminist Theory*, edited by Linda Nicholson (London: Routledge, 1997).

41. Beverley Skeggs, *Formations of Class and Gender. Becoming Respectable* (London: Sage, 1997).

42. Dorothy Smith, *The Everyday World as Problematic: A Feminist Sociology* (Milton Keynes: Open University Press, 1987), 70.

43. Nancy Fraser, "Structuralism or Pragmatics? On Discourse Theory and Feminist Politics," in *The Second Wave: A Reader in Feminist Theory*, edited by Linda Nicholson (London: Routledge, 1992/1997).

44. Michel Foucault, "Truth and Power," in *Power/Knowledge: Selected Interviews and Other Writings 1972–1977 by Michel Foucault*, edited by Colin Gordon (New York: Pantheon Books, 1980) 109–34.

45. Foucault, "Truth, Power," 118.

46. Foucault, "Truth, Power," 114.

47. Foucault, "Truth, Power," 132–33.

48. Michel Foucault, *Madness and Civilisation: A History of Insanity in the Age of Reason* (London: Tavistock, 1989).

49. Michel Foucault, *Discipline and Punish: The Birth of the Prison*, trans. Alan Sheridan (London: Penguin, 1977).
50. Michel Foucault, *The History of Sexuality Vol. 1*, trans. Robert Hurley (New York: Vintage, 1980b).
51. Foucault, "Truth, Power," 114.
52. Fraser, "Discourse Theory," 380.
53. Andermahr, Lovell and Wolkowitz, *Glossary, Feminist*, 172.
54. Anthony Elliot, "Psychoanalysis and Social Theory," in *The Blackwell Companion to Social Theory*, edited by Bryan Turner (Oxford: Blackwell, 1996), 181.
55. Fraser, "Discourse Theory," 384.
56. Samantha Ashenden, "Feminism, Postmodernism and the Sociology of Gender," in *Sociology after Postmodernism*, edited by David Owen (London: Sage, 1997).
57. Judith Butler, "Imitation and Gender Insubordination," in *The Second Wave: A Reader in Feminist Theory*, edited by Linda Nicholson (London: Routledge, 1997), 301–8.
58. Paul Connolly, "Racism and Post Modernism. Towards a Theory of Practice," in *Sociology after Postmodernism* edited by David Owen (London: Sage, 1997).
59. Ruth Frankenburg, *White Women, Race Matters. The Social Construction of Whiteness* (London: Routledge, 1993).
60. James and Busia, *Black Feminisms*.
61. bell hooks, *Killing Rage—Ending Racism* (New York: Henry Holt, 1996), 35.
62. Connell 1987 cited in Marshall, *Engendering Modernity*, 90–91.
63. Andermahr, Lovell and Wolkowitz *Glossary, Feminist*, 170.
64. Andrew Sayer, *Method in Social Science: A Realist Approach* (London: Routledge, 1992), 84 and 253.
65. Marshall, *Engendering Modernity*, 159.
66. Ashenden, "Feminism, Postmodernism," 55.
67. Sally Baden and Anne Marie Goetz, "Who Needs (Sex) When You Can Have (Gender)? Conflicting Discourses on Gender at Beijing," *Feminist Review*, no. 5 (Summer 1997): 3–25.
68. Baden and Goetz, "Sex, Gender," 6, citing Kabeer.
69. Diane Bell and Renate Klein, *Radically Speaking: Feminism Reclaimed* (London: Zed Books, 1996), 561.
70. Pierre Bourdieu, Jean-Claude Chambordon and Jean-Claude Passeron, *The Craft of Sociology. Epistemological Preliminaries* (New York: Walter de Gruyter, 1968/1991).
71. Harding, *Science, Feminism*.
72. Michel Foucault, "Truth, Power."
73. Sue Clegg, Research Methodology Teaching Materials, unpublished (Leeds Metropolitan University, 1998), 5.
74. Linda Alcoff, "Cultural Feminism versus Post Structuralism: The Identity Crisis in Feminist Theory," in *The Second Wave: A Reader in Feminist Theory*, edited by Linda Nicholson (London: Routledge, 1988/1997), 343.
75. Alcoff, "Cultural Feminism," 348.

76. Alcoff, "Cultural Feminism," 350.
77. Alcoff, "Cultural Feminism," 342.
78. Marshall, *Engendering Modernity*, 119.
79. Sherry Gorelick, "Contradictions of Feminist Methodology," in *Race, Class and Gender: Common Bonds, Different Voices*, edited by Esther Ngan-Ling Chow, E. Doris Wilkinson and Maxine B. Zinn (London: Sage, 1996).
80. Nira Yuval Davis, *Gender and Nation* (London: Sage, 1997).
81. Stanley and Wise, *Breaking Out*.
82. Yuval Davis, *Gender, Nation*.
83. Marshall, *Engendering Modernity*, 109.
84. Yuval Davis, *Gender, Nation*, 4.
85. Yuval Davis, *Gender, Nation*, 67.
86. Marshall, *Engendering Modernity*, 111.
87. Fiona Williams, "Somewhere Over the Rainbow: Universality and Diversity in Social Policy," *Social Policy Review* 4, edited by Nick Manning and Robert Page (Canterbury: Social Policy Association, 1992).
88. Bottero, "Gender, Class," 482.
89. Yuval Davis, *Gender, Nation*, 86.
90. Yuval Davis, *Gender, Nation*, 128.
91. Yuval Davis, *Gender, Nation*, 131, citing Assiter (1995).
92. I discuss in chapter 6 my reasons for deciding to use the concepts *colour* and *geographical heritage* rather than *race* and *ethnicity* to identify women students.
93. Smith, *Everyday World*.
94. Derek Layder, *New Strategies in Social Research* (Cambridge: Polity, 1993), 249.

3

Theorising Gender, Space and Time

"Up at 7am . . . sort the children out . . . leave . . . half eight . . . drive over to (college) for half nine . . . lectures . . . till half eleven . . . lunchtime . . . one o'clock . . . took the lecture . . . drive . . . to the childminder's . . . back home about quarter to four . . . start tea . . . for five . . . out of the door about quarter to six . . . start work . . . get back home about ten past ten."

Respondent Susan, chapter 10

"Time/space is now everything and therefore in danger of disappearing again into generalisations."

Researcher Log, chapter 11

Introduction

IN THIS CHAPTER I REVIEW LITERATURE concerned with conceptualising space and time in social theory. Following this I outline the key spatial/temporal concepts which underpin my own research. This chapter builds on the points raised in chapter 2. I argue that making space and time visible in the research process facilitates insight into the wider social relationships, social positions and personal actions shaping women's educational achievements. The chapter is in three parts considering: concepts of action in social theory; concepts of space and time in social theory; and spatial/temporal concepts underpinning the research. In the first two parts I review a range of authors, identifying critical issues and key concepts. In the final section, I outline the key concepts of space and time I have developed from this literature review and explain the ways in which these are applied in analysing further literature and in developing the research.

Concepts of Action in Social Theory

Ira Cohen[1] provides an overview of sociological theories of action. He traces the development of concepts of action from western philosophy of the nineteenth century. Concepts from utilitarian philosophy in particular, for example, the belief that individual actions are taken in order to *"satisfy wants or minimise loss"*[2] inform the theories of Weber and Parsons. Weber argues that actors may follow a variety of motives including *"commitment, intimidation, self-interest or habit."*[3] Actors adhere to maxims or rules which constitute *"ideal types"* of institutional organisation or *"supra individual norms of rationality."*[4] These emerge from collective social habits and traditions towards which actors orient their behaviour.[5] Parsons argues that all individual actions are rational and based on self-interest, even where they may appear irrational. Individual actors act according to their own desired ends and in the context of social values and rules.[6] A key development in social action theory was the shift from a focus on the mental processes involved in action to a focus on actual processes of enactment.[7] In the symbolic interactionist school all action is conceptualised as interaction, and the school of ethnomethodology emphasises the order of social meanings that shape action at a local level.[8] Action should be understood in its local context and frame of meaning.[9] Cohen identifies two limitations in social action theory. Firstly what significance does emotion have in relation to action? Theorists have emphasised *reason* as connecting the individual with the social world, rather than emotion. Secondly, Cohen argues that concepts of *power* are not fully theorised. These limitations are examined as follows.

Emotion and Action

Allison Jaggar argues that *emotion* has been set apart from *reason* in the development of social scientific theory. She makes the case for emotions to be considered as *intentional*,[10] *socially constructed*[11] and *active engagements*.[12] *Outlaw emotions*[13] give insight into the effects of unequal power in society. They represent the disjuncture between dominant representations of how things should be and personal experiences: *"Critical reflection on emotion is not a self-indulgent substitute for political analysis and political action. It is itself a kind of political theory and political practice, indispensable for an adequate so-*

cial theory and social transformation."[14] Analysis of personal emotion highlights areas of conflict and complexity. In addition, emotions are in themselves forms of action. Arlie Hochschild asks, *"What emotional paradox are we apparently trying to resolve in order to live the life we want to live? By what dialectical interaction between feeling rules and context is this emotional paradox produced?"*[15] The idea that action is rational and based on economic self-interest is also countered by Rosalind Edwards and Simon Duncan[16] in their research into mothers' choices in relation to mothering and paid work. They argue that such decisions are framed through *"gendered moral rationalities"* within which emotions and social values are highly significant.

Power and Action

Several writers have focused on issues of social power in relation to action. Questions are raised as to the degree to which action is either autonomous or socially determined and the ways in which wider relationships of power inform and constrain action. Jurgen Habermas[17] brings together concepts of *system, bureaucratisation* and *capitalism* from Weber and Marx in order to develop a theory of *communicative action*. His approach facilitates a deeper understanding of the power involved in the relation between structure and action. Habermas argues that the changing nature of advanced societies following from developments in industry, the city and capitalism has created a complex frame of reference for social action. He argues that there has been an increasing separation between *system* and *life-world*. The concept of *life-world* describes the way particular communities understand and order their world.[18] The *system* describes patterns of administrative organisation. Habermas argues that in *"advanced"* societies, highly complex systems have developed, including financial, legal and professional. These he terms *steering media* as they provide system integration, replacing the exchange of goods and the authority of elders as forms of social regulation.[19] Individual actors may perceive these systems as objective social facts but they carry their own *"imperatives for action."* These may conflict with social beliefs and personal understandings[20] and *"rob actors of the meaning of their own actions,"*[21] leading to feelings of fate and powerlessness.[22] The systems of social welfare (including education) may have dysfunctional effects for individual action.[23] For example, the norms of social behaviour may clash with the system's own imperatives. Clear lines of power related to the interests of powerful actors shape these systems: *"The situation to be regulated is embedded in the context of a life history and a concrete form of life; it has to be subjected to violent abstraction, not merely because it has to be subsumed under the law, but so that it can be dealt with administratively."*[24] Gendered and "racialised" divisions remain under-theorised in Habermas's work. In addition, Nick Crossley[25] has argued that one limitation of Habermas's theory of communicative action is that he, too, fails to address the significance of emotion.

Social action theory, so far reviewed, underplays the spatial/temporal dimension of action. The concepts of *system, structure, rules,* and *norms* encourage a focus on static structures rather than processes. In relation to the concept of action, the emphasis is on a fairly small range of behaviours and mental processes, in particular reasoning.[26] Theory that considers spatial/temporal concepts give more attention to the processes involved in action, as examined below.

Concepts of Space and Time in Social Theory

John Urry argues that much twentieth-century social theory has ignored the significance of the concepts of space and time as centrally meaningful. These have been theorised as neutral, and as providing an abstract and invisible context for social events. He argues, "*Social relations are irreducibly temporal . . . there are various different social times embedded within particular social structures.*"[27] Where space and time have been explicitly considered as conceptual tools within social science, the focus has been on mapping and measuring rather than on process and meaning. Echoing this, John Hassard[28] points to a sociological tradition that focuses on structure rather than *"flux and change."*

Karen Davies[29] discusses the ways that linear conceptualisations of time are associated in particular with the male priesthood and the development of Judeo/Christian religion. The idea that the world could be controlled through prayer led to the development of mechanisms to measure time and regulate prayer, despite local time systems. She argues that the development of contemporary notions of time is centrally related to the attempt to control. In addition she identifies that this process is gendered.

The awareness of the social regulation of time and space as a vehicle to extract profit has informed both Marxist and feminist theories. Thompson points out that a dominant orientation to *clock* time is a feature of industrial capitalism: "*not the task, but the value of time when reduced to money, is dominant. Time is now currency, it is not passed but spent.*"[30] The regulation of space and time involves "*a new self discipline related to work pattern.*"[31] The transition from task oriented to clock oriented time during industrialisation features the factory owners teaching workers the importance of clock time. Struggles over time and its value shape industrial action.[32] The separation of home from workplace, also a feature of industrialisation, relates to the drive for greater production and profit. Peter Williams argues that this separation involves the reconceptualisation of the spaces of home and of work.[33] Feminists have been centrally concerned with the regulation of space and time as it shapes women's experiences. The themes of women's labour time, of public and private spheres, of the inclusion/exclusion of women and of the impact of *others' time*[34] on women's lives have been at the heart of feminist debate (see chapter 4).

In relation to concepts of space in social theory, Marxist geographers began to insist that geographical boundaries being mapped within the discipline were not the consequence of mere spatial systems and spatial distributions. They were the outcome of social relationships over time and influenced by relationships of power.[35] Developing from an awareness of places as social outcomes, radical geographers (for example, Doreen Massey) began to explore the way in which spatial arrangements in their turn influenced social outcomes. They argue that the spatial should be viewed in terms of its social construction and the social in terms of its spatial construction.[36]

In terms of the social analysis of time, Hassard[37] traces the development of the concept of *qualitative time*, discussing theorists who have explored the social meanings attributed to time span and the inadequacy of astronomical models in social science. For example, Elliot Jaques[38] argues that *"life is different from physics."* Our sense of time involves *"memories in the present of the past, expectations and desires in the present of the future."* As early as 1937, Pitkim Sorokin and Robert Merton[39] discuss the limits of the astronomical model of time and write of *ontological time*—the time of *being*. They argue that local time systems relate to local experience and are inherently qualitative. For example some communities have *told* time by the market day, or by particular agricultural cycles. Sorokin and Merton trace the process whereby one-dimensional astronomical time substituted for local time systems[40] and relate this to the need to *"regulate the collective."* Hence, they argue that the origins of astronomical time are social. Jaques[41] rejects the attempt to draw on astronomical models of space/time including relativity theory in order to understand social time.

The sociology of time involves exploring dichotomous explanations of time as subjective/objective, cyclical/linear, and clock based/event based. Hassard[42] argues that within social science time can be used as:

- a social factor (either as a resource or as a source of meaning);
- a causal link (in this time, that happened);
- a quantitative measure;
- a qualitative measure.

Barbara Adam[43] argues that it is not possible to make visible individual agency and political action unless we understand the significance of time as a social dimension. Time has a multiplicity of meanings and is experienced and organised in different ways. It is socialised around benchmarks that might suit some interests more than others[44] and is a medium of power and an outcome of power relations.

I will now move on to discuss three concepts within social theory that have involved space and time as central themes. These are the concepts of *structuration*,[45] *the production of space*[46] and *habitus*.[47] Following this I will discuss in more detail some of the dualist thinking associated with the concepts of space and time and the implications of this. Rather than viewing the human/social as separate from its spatial/temporal context, critics argue that the spatial/temporal penetrates the social sphere, for example through human memory, through culture, through artefacts, through symbols and signs, through technology and through spatial and temporal practices.

Structuration Theory

Anthony Giddens draws on the time-geography of Hagerstrand to develop *"an ontological framework for the study of human social activities . . . a conceptual investigation of the nature of human action, social institutions and the interrelation between action and institutions."*[48] He emphasises the duality of agency and structure, *"Locally situated though they may be, all practices also contribute to the production and reproduction of systemic relations and structural relations."*[49] Social science has been hampered by dualist theories of social structure and agency. Structures have been conceptualised as overly determinist and human agency has disappeared from analysis (for example, in some economic theory). Alternatively, individuals have been theorised as disconnected from the social, and as rooted in their own personal history and psychology (for example, in some psychological theory). Giddens's theory of structuration attempts *"to understand both how actors are at the same time the creators of social systems, yet created by them."*[50] Although structures of power, such as class, are significant, they provide *"no overall motor of change."*[51] There are different planes of temporality: *duree*—daily experience; *dasein*—the life course; and *longue durée*—the time of social institutions. All of these bear upon social events. In order to further understanding of structuration, Giddens develops a very detailed typology of space/time which is reviewed by Urry and is *"concerned with showing how life processes are linked to the long duree of institutions."*[52] Integral to Giddens's perspective is the visibility of spatial/temporal factors: *"I have sought to show how time and space might be brought into the core of social theory and . . . how time-space is handled within differing types of social system, thereby entering into their constitution."*[53]

The body of concepts Giddens develops is intended to deepen understanding of the spatial and temporal dimensions in the relationship between the social and the individual. A reading of Hassard[54] shows that several theorists have previously attempted to classify or "map" the concept of time (see for example the work of Georges Gurvitch).[55] In my own research I have not attempted to develop a detailed classification of spatial/temporal concepts as above but have borrowed from these ideas (see below, "Spatial/Temporal Concepts Underpinning the Research" and chapter 11).

The Production of Space—Henri Lefebvre

John Urry[56] argues that Giddens gives insufficient attention to the concepts of *space* and *place*, that is, to the way changes over time have affected the specific geographies of place. One aspect of Giddens's emphasis on internationalisation and space/time (through considering the development of financial, communication, and other technological relations) is an implied assumption that the significance of specific places has decreased. However, *"places and their locale . . . refer to a part of social space in which people live, work and socialise and the interaction over time gives places their distinctiveness."*[57] A theorisation of *space* is therefore required. Henri Lefebvre[58] focuses on processes of spatial formation and their significance in terms of power and inequality. From a Marxist position, Lefebvre argues that space is socially produced. This involves the dialectical interaction of three elements:

- *spatial practice*: for example spatial formations of production, reproduction, habitation and so forth;
- *representations of space*: for example the signs, codes, knowledge and artefacts related to space;
- *representational space*: for example art, philosophy and other activities, which have a transformative potential in terms of both practice and representations.

Lefebvre argues that *"space is not neutral or passive geometry"*[59] but is a site of struggle. His conceptualisation allows understandings of the way major structures of power such as class penetrate spatial formation and representation. Space is not merely context but shapes and is shaped by interaction. In his analysis, spatial process and *place* have significance as carriers of power. *Space* is both imagined and understood. *Place* is a product of human activity and carries the residue of the values informing that activity. I have drawn heavily on the above concepts in my own research and a more detailed discussion of each is to be found below in "Spatial/Temporal Concepts Underpinning the Research."

Field, Habitus and Body

Although both Giddens and Lefebvre deepen understanding of the relationship between the individual and the social through addressing spatial and temporal processes and meanings, there remains a problem. The dualism of structure and agency is addressed and yet the dualism of the individual and the social remains. The individual is represented as separate from the sphere s/he occupies. For example, Giddens argues that social actors are both recursive and reflexive—recreating activities yet constantly monitoring them.[60] This idea represents individuals as at a distance from the social world rather than as central to it.

Both Thrift and Connolly discuss the work of Pierre Bourdieu who provides an understanding of the position of the individual in relation to the social. He uses spatial metaphors in distinguishing between *"field . . . of social interaction . . . a network of objective relations redefined by agents—based in certain forms of power . . . Habitus . . . an embodied state . . . a kind of embodied unconscious."*[61] "In essence the habitus is defined and constituted within particular fields as people come to learn about and internalise their position with struggles over particular forms of capital."[62] Bourdieu conceptualises capital as *economic, cultural, social* and *symbolic*.[63] *Habitus* then is a state of being which has conscious and non-conscious parts and is related to the field where historical processes come to fruition. Shirley Ardener,[64] drawing on Bourdieu, uses the term *"schemes of perception"* to account for the different ways women conceptualise their social world.

A number of theorists, building on ideas concerned with the complexity of *being* in the social world, have developed a more complex theorisation of the body. The body, including thoughts, emotions, and the physical, is theorised as an expression of the *"materiality of connections"* (including social). Bodies are *"vehicles of perception"* and the *"object perceived."* Bodies are not fixed and natural but *"located in time and space."*[65] *Being* is not easily separable from context, neither are things easily separable from the body. The example Thrift gives is that of the stick someone may use to scratch his or her body. At what stage does this object become an extension of the body? The body is engaged in using and creating value out of things in the world. Citing Merleau-Ponty, Thrift argues that *"bodies unite past, present and future . . . take possession of time."*[66]

Feminists and disability theorists (such as Ruth Butler and Sophie Bowlby) have been at the forefront of theorising the social significance of space, the body and emotion, *". . . treating the experience of the body as the outcome of a reflexive relationship between bodily materiality and social process. The body is not a passive and fixed 'fact' onto which social relations are mapped, nor can what seem to be physical experiences of the body simply be accepted as 'facts' which are prior to or determinative of social relations."*[67] They argue that a reconceptualisation of the body enables us to unpick some of the dualist theorising about the individual and the social sphere. For example, there are difficulties in giving a full account of the experiences of disability and of ageing which are constrained neither by the limitations of the medical or social models of age and disability. The former ignores the social construction of age/disability, but the latter may marginalise personal experiences of pain associated with impairment.[68] Through reconceptualising the body we are able to see the body as social. As a result, bodily experiences *and* their relationship to the social world remain visible.

Dualisms of Space and Time

It is feminist writers who have drawn attention to the gendered dualism inherent in contemporary conceptualisations of space and time. Barbara Adam, as introduced above, draws attention to the fact that social research has drawn on limited understandings of time as based on a Newtonian model. In this perspective time is considered as linear and space as contextual. Time is considered a measure of things and events. This perspective is particularly reflected in positions that theorise social events as progressive and linear.[69] Doreen Massey shares Adam's critique. She points out that separating the analysis of space and time has led to time being theorised as the dynamic concept, concerned with social change, and space as the static concept.[70] Time has been associated with ideas about action, progress, transformation and masculinity and space with ideas about emptiness, stasis, context and femininity.

Adam also challenges an assumption embedded in social science that time is either *social* or *natural*. Central to contemporary thinking about time is the idea that in the social sphere time is linear and progressive, whereas in the natural sphere (associated with seasonal, bodily, agricultural processes and so forth) time is cyclical. This dualist thinking has had major implications for the way that the experience of particular social groups has been theorised. For example, where mainstream social science has considered the position of women in general, men and women living in non-industrialised communities and groups defined in terms of racial or ethnic characteristics, ideas about cyclical time come to the fore. These groups are often presented in terms of their closeness to nature. Adam forcefully argues that it is a false duality to distinguish between social and natural time in this way. She argues that *"human time includes the time of nature . . . human time is the time of nature."*[71]

Central to feminist thinking has been the attempt to draw the concepts of space and time closer together and to create space for qualitative research approaches whereby ideas about *becoming* and *being* remain visible. For example Adam draws on ideas from the new biology, where organisms are conceptualised in terms of their *rhythmicity*. She draws on the concept of living things as temporal beings, "*practising centres of action rather than perpetrators of fixed behaviour.*"[72] Integral to her argument is the concept of multiple rather than single time spans shaping experience, for example the times of the body, senses, environment and industry. Massey argues that there is a need to conceive of the concepts of space and time as interdependent and equally involved in social process. Both Adam and Massey show the potential for feminism of a more complex theorising of space and time in addressing issues of power and inequality.

Spatial/Temporal Concepts Underpinning the Research

Summary of Concepts from the Literature

The array of overlapping analysis so far discussed provides a range of conceptual tools that sit uneasily together. The concept of *economic rational man* in social action theory is contested. The concepts of *emotion* and *power* are insufficiently addressed, but even where they are, the invisibility of spatial/temporal factors weakens our understanding of processes of social action. Giddens develops concepts concerned with identifying the spatial/temporal dimension of social relations. He focuses on the development of new technologies and forms of communication and shows the way these are involved in processes of structuration. As John Urry[73] argues, he gives insufficient attention to the significance of the social history of place in his theory. Gillian Rose has argued that other aspects of experience are excluded from analysis in time geography. She argues that the attempt to map and classify social relations in objectified ways has severe limitations: "*The search for totality in time geography involves constructs of masculinity.*" She asks whether time geography is sufficient to "*speak of women's subjectivity,*" for example, "*The emotional, the passionate, the disruptive and the feelings of relations with others.*"[74] Nevertheless, from Giddens we learn of the inseparability of space and time conceptually, and the importance of making space and time visible in order to understand the relationship between structure and agency. His concepts also increase the visibility of social relations that are *stretched* through space and time, for example through processes of internationalisation and institution. From Lefebvre we learn more sophisticated ways of theorising space (as both practice, and representation). However, it seems time is insufficiently theorised in his work. I had difficulty for instance in understanding the difference between the concepts of *representations of space* and *representational space*. At what stage does *representational space,* for example, acts of creation, become *representation,* for example, signs and codes (see below)? In both Giddens and Lefebvre the domestic sphere and gendered relations are given insufficient consideration.

From Bourdieu we begin to further break down the dualism involved in theorising the individual and the social. The concept of *habitus* shows the way the individual is integrally related to the social. New conceptualisations of the body serve a similar purpose. Adam's conceptualisations of individuals as *centres of action* and of *multiple time spans* are relevant here.[75] As Jaques argues,[76] in order to develop an adequate sociology of time it is necessary to draw on many different conceptualisations and related theories. Below I outline the key spatial/temporal concepts I use in my own research. I have drawn out and combined conceptualisations that make most sense to me and fit comfortably with my research. These are threefold: *spatial/temporal practices*; *spatial/temporal representations*; and *centres of action.*

Spatial/Temporal Practices

This concept is drawn from combining the work of Lefebvre,[77] Adam[78] and Davies.[79] It gives visibility to the dominant ways in which space and time are organised, for example, public/private spheres and the regulation of labour time in and across both those spheres. It draws on Davies's discussion of *temporal rationalities* (see below) as well as Lefebvre's concept of *spatial practice.* It enables us to see the way *others' time*[80] and space shape our own experience. The concept of *spatial practice* "embraces production and reproduction, and the particular locations and spatial sets of characteristics of each formation."[81] Lefebvre writes from a Marxist standpoint, being concerned to reveal "*spatial practice under neo-capitalism . . . daily reality (daily routine) and urban reality (the routes and networks which link up the places set aside for work, 'private' life and leisure).*"[82] He argues that the association of these realities is paradoxical involving "*extreme separation between the places it links together.*" Capitalism involves the "*manipulation of space.*"[83] Clearly sex-gender systems also involve spatial production.[84] Shirley Ardener uses the concepts of *ground rules* and *social maps* to explore similar terrain: "*Societies have generated their own rules, culturally determined, for making boundaries on the ground, and have divided the social into spheres, levels and territories with invisible fences and platforms to be scaled by abstract ladders and crossed by intangible bridges with as much trepidation or exultation as on a plank over a raging torrent.*"[85] The concept *social map* describes the structural/spatial relations in particular societies, for example, hierarchies, ranking and kinship systems, public and private domains. Ardener argues that *"behaviour and space are mutually dependent."*[86]

The concept of *spatial practice* relates closely to the concept of *temporal rationality* and *dominant temporal consciousness* discussed by Davies.[87] Rationalities related to time are involved in social and spatial practice. Davies discusses these as fourfold: *"technical economic . . . scientific . . . earner (and) . . . nurturing and responsibility."*[88] These rationalities are accorded different social value in relation to different areas of practice. For example, in the sphere of paid care work, the first three take priority at the expense of the fourth. In the home and in relationships, the latter may take more priority. These rationalities are forms of knowledge emerging from former social and spatial practices and in this sense they may more closely relate to Lefebvre's concept of *spatial representations*. However, as they are modus operandi for dominant social and spatial practices, I have brought the concepts together.

In my own research I use the concept *spatial/temporal practices* to identify *snapshots* of the relations women are involved with in the sphere of paid work, home, leisure/community and higher education. By *snapshots* I mean to convey an image of the web of relations women are involved with at specific times and places in their academic career. I discuss women's social position, paid work, housing and household and leisure/community. These *snapshots* are discussed in chapters 7 and 9.

Spatial/Temporal Representations

This concept draws on Lefebvre's concept of *spatial representations*, but makes the temporal dimension explicit. It gives visibility to the inanimate, artefacts, symbols, signs and codes of meaning and knowledge systems. The way places and things emerge and the lines of power involved become more visible.

Clearly there is considerable overlap between the two concepts of *spatial practices* and *spatial representations*, but they nevertheless provide useful analytic tools. The concept of *spatial representations* gives visibility to the emergence of places and things. The concept *place* is used in a broad sense to include not just physical places but the places of knowledge and artefacts. I have adapted Lefebvre's concept by drawing on the work of Adam, Davies and Hassard. Time is a highly complex and contested concept.[89] Different meanings associated with time are associated with specific places. For example, the places of employer/employee relations are particularly associated with the concept of clock time, which is quantifiable and has a monetary value. In addition, places themselves emerge in a temporal context and represent residues of former practice and associated social value. The functions of particular places relate to the patterns of relations giving rise to them. For example, places may be inclusive of particular categories of people, things and ideas or may be exclusive, depending on the times they are used.[90]

In this research I draw on the concept of *landscape*[91] to explore dominant *spatial/temporal representations* shaping women students' lives. Cara Aitchison discusses the concept of *landscape* from new cultural geography as *"not fixed but . . . in a constant state of transition as a result of continuous dialectical struggles of power among and between the diversity of landscape providers."*[92] Citing Daniels and Cosgrove, she argues, "*Rather than being a physical or objective reality, landscape is a cultural image, a pictorial way of representing, structuring and symbolising surroundings."*[93] Time is also central to the landscape. Adam develops the concept of *timescape* to explore *"complex temporalities of contextual being."*[94] Landscapes mean different things to different individuals who experience them. In this research, landscapes are developed from a variety of elements, which separate and connect different parts, for example different spheres of experience. The concept of *landscape* provides a useful vehicle for considering the *socially dominant* meanings associated with the different spaces women occupy. In chapters 8 and 9 I have developed three *landscapes* from women's accounts of their experience. The *physical landscape* includes concrete spatial features, for example, architecture and geographical zoning. The *social landscape* includes relationships with others in family, kinship and friendship networks. The *emotional landscape* gives visibility to women's feelings and areas of *emotional dissonance*, as they are associated with the different places they experience. It includes areas where life experience is at odds or discordant in relation to dominant representations. They may be areas associated with emotional unease, material insecurity or low social status. The concept of *emotional dissonance* is developed from the work of Allison Jaggar[95] and Arlie Hochschild,[96] whose respective concepts of *outlaw emotions* and *emotional paradox* give insight into the effects of unequal power in society. They represent the disjuncture between dominant representations and personal experiences. In the case of each landscape, associated concepts of time are considered. These are widely overlapping landscapes. The overlaps in themselves draw attention to the connections between the physical, social and emotional in women's lives. They are not separately but simultaneously lived through.

Centres of Action

This concept from Barbara Adam, together with Bourdieu's concept of habitus and reconceptualisations of the body discussed earlier, provide the potential for a rich analysis of individual experience in its material context. Integral to it is the concept of *multiple time spans* that shape people's lives. Women's struggle for multifunctional space may also become visible.[97] This concept also has affinities with Lefebvre's third concept, that of *representational space* because it allows consideration of human behaviour in a wide sense, including the spheres of emotion and bodily needs and desires. Lefebvre attempts to convey the concept of action without disconnecting it from context: "Representational space: space as directly lived through its associated images and symbols, and hence the space of 'inhabitants' and 'users' . . . This is the dominated and hence passively experienced space, which the imagination seeks to change and appropriate. It overlays physical space, making symbolic use of its objects."[98] In an almost literary way he attempts to show the way that human action is embedded in spatial structures: "*representational space is alive: it speaks, it has an affective kernel or centre: Ego, bed, bedroom. Dwelling, house; or square, church, graveyard. It embraces the loci of passion, of action and of lived situations, and this immediately implies time.*"[99]

In Lefebvre's conceptual triad, s*patial practice* is *perceived space, spatial representation* is *conceived space* and *representational space* is *lived space*.[100] The complexity of the relationship between social, spatial and system arrangements and individual action is again conveyed through the concept of *landscape* (see above): "*The power of the landscape does not derive from the fact that it offers itself as a spectacle, but rather from the fact that as mirror and mirage, it presents any susceptible viewer with an image at once true and false of creative capacity which the subject (or ego) is able, during a moment of marvellous self deception, to claim as his own.*"[101] The *landscape* emerges from social life. It directs the actor's gaze and the actor can also engage with and transform the landscape. The *"moment of marvellous self deception"* (see above quotation) is presumably the moment when actors feel they have more power to change things than is the reality. Although Lefebvre's language is very difficult to understand, it is clear that concepts of action in his theory are temporally and spatially situated. The focus on reason, motivation and emotion in social action theory is broadened through an explicit focus on the body and production. The influences on him here are Marxist, *"for the body indeed unites cyclical and linear, combining the cycles of time, need and desire with the linearities of gesture, perambulation, prehension and the manipulation of things—the handling of both material and abstract tools.*"[102]

How do we analyse the connections between structure and action (in social action theory) and between *spatial practices, representations* and *representational space*? "*How exactly the laws of space and of its dualities (symmetries/asymmetries, demarcation/orientation etc.) chime with the laws of rhythmic movement (regularity, diffusion, interpenetration) is a question to which we do not have the answer."*[103] Both Adam and Lefebvre highlight the potential of the concept of *rhythm* for the analysis of action. In conceptualising beings as *centres of action*, Adam demonstrates the potential of *rhythm* as a concept in social research. She argues that this gives visibility to the complex of conditions shaping human experience. The concept focuses on action, but requires that all the contexts of action are given visibility and that action is spatially and temporally situated. She argues that individual actors live temporally and through different rhythms. These include the rhythms of their bodies (heartbeats, menstrual cycles, life-span), the rhythms of their households, of agricultural cycles, of paid employment and industrial time, of historical and institutional time.[104] Human action may be conceptualised as *"symphonies in rhythm"*[105] and as *"the expression of different levels of our being."*[106] *"As time, timing, tempo and temporality we can recognise some of the complexity of that which is ultimately individable ... Through its rhythmicity life becomes predictable. Thus the focus on time helps us to see the invisible."*[107] Adam stresses that approaches that give visibility to space and time avoid the researcher needing to make *"unacceptable choices between biological and social determinants . . . between realism and relativism . . . between meta-narratives and particularism."*[108]

In order to operationalise the concept *centres of action* in my own research, I have developed and drawn on further concepts. These concepts arise from the above arguments of Adam and Lefebvre in relation to *rhythm*. From there I have tried to identify concepts which focus on movement through both space and tempo. By addressing both the spatial and temporal aspects of action I hope to gain fuller insight into the actions women take in order to integrate higher education into their lives. The key concepts I have developed to explore women students as *centres of action* are:

- *Pathways*. How do women move through the different landscapes that they experience, physical, social and emotional?
- *Patterns of spatial use*. How do women move between the interconnecting landscapes they experience?
- *Rhythm*. What are the multiple rhythms shaping women's experience? How is the rhythm of higher education accommodated? In pursuing higher education how do women negotiate *others' time*[109] and time for other things?
- *Place for studies*. What places do women create for studies at a physical, social and emotional level?

The analysis of women students as *centres of action* is contained in chapter 10.

Conclusion

The inseparability of the concepts of space and time is stressed in my research. Often it is hard to separate the concepts as analytic tools. Space and time are experienced simultaneously. In addition, our conceptualisations relate to our social position as researchers. The unity of space/time as a concept is evident in other cultures such as the Runa of Quechua in the Andes. The word *Pacha* in their language signifies both space and time: *"Time is neither past, present, nor future but eternally being and space is the physical/cultural realm in the state of existence."*[110] Sarah Lund Skar argues that practically and conceptually it is impossible to separate experience as spatial from experience as temporal. It may be possible to analyse the different meanings people ascribe to these concepts and it is essential that the researcher's meaning is both contextualised and clarified. Below I summarise the core spatial/temporal concepts that underpin the research and show where they inform the study.

TABLE 3.1
Core Spatial/Temporal Concepts: Location

Core Concepts	Research Questions	Research Methodology	Data Presentation and Analysis
—Spatial/temporal practices	—Segregation/ separation —Public/private —Others' time	—Methodology	—Snapshots of women's experience
Chapter 3	Chapters 4 and 5	Chapter 6	Chapters 7 and 9
—Spatial/temporal representations	—Space as process, not object —Codings of social division	—Epistemology	—Landscapes of women's experience
Chapter 3	Chapters 4 and 5	Chapter 6	Chapters 8 and 9
—Centres of action	—Women's feelings and conceptualisations —Women's actions for change	—Methods	—Pathways to and through higher education —Patterns of spatial use —Rhythm of studies —Place for studies
Chapter 3	Chapters 4 and 5	Chapter 6	Chapter 10

Notes

1. Ira J. Cohen, "Theories of Action and Praxis," in *The Blackwell Companion to Social Theory*, edited by Bryan S. Turner (Oxford: Blackwell, 1996).
2. Cohen, "Action, Praxis," 111.
3. Cohen, "Action, Praxis," 115. A social relationship is "*when several actors mutually orient the meaning of their actions*" (Cohen, "Action, Praxis," 114).
4. Cohen, "Action, Praxis," 115.
5. Cohen, "Action, Praxis," 113.
6. Cohen, "Action, Praxis," 118.
7. Cohen, "Action, Praxis," 121.
8. Cohen, "Action, Praxis," 128.
9. See Nigel Thrift, *Spatial Formations* (London: Sage, 1996) for a fuller discussion of both schools of thought.
10. Allison M. Jaggar, "Love and Knowledge: Emotion in Feminist Epistemology," in *Women, Knowledge and Reality, Explorations in Feminist Philosophy*, edited by Ann Garry and Marilyn Pearsall (London: Routledge, 1996), 69.
11. Jaggar, "Love, Knowledge," 171.
12. Jaggar, "Love, Knowledge," 172.
13. Jaggar, "Love, Knowledge," 181.
14. Jaggar, "Love, Knowledge," 185.
15. Arlie Hochschild, "The Sociology of Emotion as a Way of Seeing," in *Emotions in Social Life. Critical Themes and Contemporary Issues*, edited by Gillian Bendelow and Simon J. Williams (London: Routledge, 1998), 11.
16. Rosalind Edwards and Simon Duncan, "Supporting the Family: Lone Mothers, Paid Work and the Underclass Debate," *Critical Social Policy* 53, no. 17 (November 1997): 29–49.
17. Jurgen Habermas, *Volume Two—Lifeworld and System: The Theory of Communicative Action. The Critique of Functionalist Reason* (Cambridge: Polity, 1981/1987).
18. Habermas, *Lifeworld, System*, 124.
19. Habermas, *Lifeworld, System*, 167–71, for example: "*law replaces orality,*" possessions rather than birth provide authority and "*citizenship, not kinship, [becomes] the fiction of [social] membership.*"
20. Habermas, *Lifeworld, System*, 185.
21. Habermas, *Lifeworld, System*, 302.
22. Habermas, *Lifeworld, System*, 325.
23. Habermas, *Lifeworld, System*, 363.
24. Habermas, *Lifeworld, System*, 363.
25. Nick Crossley, "Emotion and Communicative Action: Habermas, Linguistic Philosophy and Existentialism," in *Emotions in Social Life. Critical Themes and Contemporary Issues*, edited by Gillian Bendelow and Simon J. Williams (London: Routledge, 1998), 21.
26. John Urry, "Sociology of Time and Space," in *The Blackwell Companion to Social Theory*, edited by Bryan S. Turner (Oxford: Blackwell, 1996).

27. John Urry "Sociology, Time," 370.
28. John Hassard, *The Sociology of Time* (London: Macmillan, 1990), 1.
29. Karen Davies, *Women and Time: The Weaving of the Strands of Everyday Life* (Aldershot: Avebury, 1990), 20–21.
30. Cited in Davies, *Women, Time*, 26.
31. Davies, *Women, Time*, 29.
32. Davies, *Women, Time*, 30.
33. Peter Williams, "Constituting Class and Gender. A Social History of the Home 1700–1901," in *Class and Space: The Making of Urban Society*, edited by Nigel Thrift and Peter Williams (London: Routledge and Keegan Paul, 1987), 202.
34. Davies, *Women, Time*, refers to *other's time* in the singular. I have chosen to use the plural *others' time*.
35. Doreen Massey, *Space, Place and Gender* (Cambridge: Polity Press, 1994).
36. John Allen and Doreen Massey, *Geographical Worlds* (Oxford: Oxford University Press, 1995), 55.
37. Hassard, *Sociology, Time*.
38. Elliot Jaques, "The Enigma of Time," in *The Sociology of Time*, edited by John Hassard (London: Macmillan, 1982/1990), 22.
39. Pitrim Sorokin and Robert Merton, "Social-Time: A Methodological and Functional Analysis," in *The Sociology of Time*, edited by John Hassard (London: Macmillan, 1937/1990).
40. Sorokin and Merton, "Social-Time," 66.
41. Jaques, "Enigma, Time," 30.
42. Hassard, *Sociology, Time*, 2.
43. Barbara Adam, *Timewatch: The Social Analysis of Time* (Cambridge: Polity Press, 1995), 44.
44. Adam, *Timewatch*, 61.
45. Anthony Giddens, "Structuration Theory: Past, Present and Future," in *Giddens' Theory of Structuration: A Critical Appreciation*, edited by Christopher Bryant and David Jary (London: Routledge, 1991).
46. Henri Lefebvre, *The Production of Space* (Oxford: Blackwell, 1991).
47. Bourdieu in Thrift, *Spatial Formations*.
48. Giddens, "Structuration Theory," 201.
49. Giddens, "Structuration Theory," 130.
50. Giddens, "Structuration Theory," 204.
51. Giddens, "Structuration Theory," 14.
52. John Urry, "Time and Space in Giddens' Social Theory," in *Giddens' Theory of Structuration: A Critical Appreciation*, edited by Christopher Bryant and David Jary (London: Routledge, 1991), 164–67.
53. Giddens, "Structuration Theory," 205.
54. Hassard, *Sociology, Time*.
55. Georges Gurvitch, "The Problem of Time," in Hassard, *Sociology, Time*, 69–71.
56. Urry, "Time, Space," 167.
57. Allen and Massey, *Geographical Worlds*, 55.
58. Lefebvre, *Production, Space*.

59. Cited in Urry, "Sociology, Time," 391.
60. Cited in David Sibley, *Geographies of Exclusion* (London: Routledge, 1995).
61. Thrift, *Spatial Formations*, 14.
62. Paul Connolly, "Racism and Post Modernism: Towards a Theory of Practice," in *Sociology after Postmodernism*, edited by David Owen (London: Sage, 1997), 73.
63. See Beverley Skeggs, *Formations of Class and Gender: Becoming Respectable* (London: Sage, 1997) for a useful discussion of Bourdieu's concept of capital.
64. Shirley Ardener, ed., *Women and Space: Ground Rules and Social Maps* (Oxford: Berg, 1993), 5.
65. Thrift, *Spatial Formations*, 13.
66. Thrift, *Spatial Formations*, 38.
67. Ruth Butler and Sophie Bowlby, "Bodies and Spaces: An Exploration of Disabled Peoples' Experiences of Public Space," *Environment and Planning. D: Society and Space*, no. 15 (1997): 411–33, 430.
68. Jenny Morris, *Pride against Prejudice. Transforming Attitudes to Disability* (London: Women's Press, 1991).
69. Barbara Adam, *Time and Social Theory* (Cambridge: Polity, 1990), 50 and 55.
70. Massey, *Space, Place*, 255.
71. Adam, *Time, Theory*, 155.
72. Adam, *Time, Theory*, 66.
73. Urry, "Time, Space."
74. Gillian Rose, *Feminism and Geography: The Limits of Geographical Knowledge* (Cambridge: Polity, 1993), 28.
75. Adam, *Time, Theory*.
76. Jaques, "Enigma, Time."
77. Lefebvre, *Production, Space*.
78. Adam, *Time, Theory*; Adam, *Timewatch*.
79. Davies, *Women, Time*.
80. Davies, *Women, Time*.
81. Lefebvre, *Production, Space*, 33.
82. Lefebvre, *Production, Space*, 38.
83. Lefebvre, *Production, Space*, 62.
84. Gayle Rubin, "The Traffic in Women: Notes on the 'Political Economy' of Sex," in *The Second Wave: A Reader in Feminist Theory*, edited by Linda Nicholson (London: Routledge, 1975/1997).
85. Ardener, *Women, Space*, 11–12.
86. Ardener, *Women, Space*, 12.
87. Davies, *Women, Time*, 9.
88. Davies, *Women, Time*, 106–10, where these concepts are fully reviewed.
89. Hassard, *Sociology, Time*, 17–18.
90. Sibley, *Geographies, Exclusion*.
91. Lefebvre, *Production, Space*, 189.
92. Cara Aitchison, "New Cultural Geographies: The Spatiality of Leisure, Gender and Sexuality," *Leisure Studies* 18 (1999): 19-39, 29.
93. Aitchison, "Cultural Geographies," 31.

94. Barbara Adam, *Timescapes of Modernity: The Environment and Invisible Hazards* (London: Routledge, 1998), 11 and 19. Adam focuses on the environment and sustainability and argues that the imposition of industrial time is a *"powerful conceptual block"* (9), which narrows our perceptions (1). Actors feel pride in the *"dominant timescape,"* and in contemporary society's dominant (clock based) ideas about time facilitate exploitation of the physical landscape.

95. Jaggar, "Love, Knowledge," 181.

96. Hochschild, "Sociology, Emotion," 11.

97. Davina Cooper, "Regard between Strangers: Diversity, Equality and the Reconstruction of Public Space," *Critical Social Policy* 18, no. 4 (November 1998), 465–92.

98. Lefebvre, *Production, Space*, 39.

99. Lefebvre, *Production, Space*, 42.

100. Lefebvre, *Production, Space*, 39.

101. Lefebvre, *Production, Space*, 189.

102. Lefebvre, *Production, Space*, 203.

103. Lefebvre, *Production, Space*, 207.

104. Adam, *Time, Theory*, 166.

105. Adam, *Time, Theory*, 74.

106. Adam, *Time, Theory*, 161.

107. Adam, *Time, Theory*, 169.

108. Adam, *Timescapes, Modernity*, 6.

109. Davies, *Women, Time*.

110. Sarah Lund Skar, "Andean Women and the Concept of Space/Time," in *Women and Space: Ground Rules and Social Maps*, edited by Shirley Ardener (Oxford: Berg, 1993), 32.

4

The Spatial and Temporal Relations of Women's Everyday Lives

"I've found that I'm one of these that if I've a piece of work to do and the house want tidied, the housework would have to be done first because . . . with a tidy house I always felt relaxed and I would light scented candles all around the house so maybe it smells nice and make sure the place is Hoovered up and tidy and I then feel more relaxed to sit down and do a piece of work."

Respondent Zandra, chapter 8

"[My son's] poorly . . . I've written recently about body times interfering with external regulation! Well here it is. I've been sitting . . . ploughing through tasks . . . notes, cards, references; money for [fees]; book to order; chucking and sorting. Midway went to the GP . . . my son vomited outside the health centre."

Researcher Log, chapter 11

Introduction

THE FOLLOWING TWO CHAPTERS provide a further literature review in order to identify key critical questions about space and time that feminist and other researchers have raised and to relate these to the spatial/temporal concepts so far identified. The aim is to identify research questions concerning women's pursuit of higher education in the context of their other activities. At this stage I consider existing feminist research in the spheres of home, paid work and leisure/community and explore the spatial/temporal relations shaping women's everyday lives and thus informing their relationship to higher education. Throughout the research I discuss women's community participation (for example, voluntary work) in relation to their experience of leisure. The ambiguity of the concept of leisure means that for many women, leisure is not conceptualised as time for self, but as time for others. The boundaries between leisure and community participation are blurred. This chapter is in three parts, considering literature concerned with: gender and spatial/temporal practices; gender and spatial/temporal representations; and women as centres of action.

Gender and Spatial/Temporal Practices

Segregation/Separation

In relation to the concept of segregation, Daphne Spain[1] examines a range of anthropological and historical material drawing from the experience of women in both the southern and northern hemispheres. She shows that the segregation of women in the dwelling, within educational processes and within the workforce all link closely to their stratification. She argues that the stratification of women in this context refers to their status in relationship to the ownership of property, their political and their labour market rights. She particularly stresses the significance of segregated education. This is the sector where the knowledge, skills and understanding deemed necessary for full participation in society are imparted. If women are excluded from this process, their lower status is reinforced. She argues that the theme of spatial segregation is relevant within Marxist theory, in terms of considering women's relationship to the means of production, and in Weberian theory, in terms of considering women's distance from social estimations of honour. Spain's work takes a macro view of women's experiences both historically and globally. In terms of U.S. women's experience, she argues that women's status has increased as spatial segregation has reduced, using the measures of status referred to above.

Spain's work is important but there are potential difficulties. Firstly, she herself cites exceptions to the general trends she identifies. Secondly, the concept of women's social stratification she draws on does not fully consider issues related to violence against women, for example its incidence and form. Even where space is not formally segregated women's movement is regulated through the use of violence and its threat.[2] Thirdly, the ways women themselves consider issues of spatial segregation and social honour vary considerably across communities and cultures. Questions arise as to the ways women themselves are involved in the construction of space and use separate spaces in order to improve their own social positioning and life chances.[3]

Public/Private

Peter Williams explores the shifts in the social construction of house and home in the period 1700–1901 in the northern hemisphere. Shifting relations of production and reproduction led to the growth of the home being conceptualised as *private*. The withdrawal of production from the home is associated with the development of rituals of correct behaviour and sexuality within the home and shifting ideas about ownership, privacy and rights of control.[4] Although designated a private sphere, public temporal practices regulate home time.[5] Feminists have identified the way the conceptualisation of *public/private* spheres involves the association of certain behaviours and roles with certain spaces. Patricia Hill Collins[6] argues that such conceptualisations decrease the visibility of black women's lives. Where women fail to conform to such boundaries either their visibility is decreased or they are considered atypical. The concept of *public/private* has strengthened insight into some aspects of women's oppression but has also at other times generated stereotypes. Fascinating accounts have been written of what occurs when individuals and groups attempt to transgress spatial boundaries in relation to violence, childcare, paid work and leisure.[7] Concepts of *public/private* also mark boundaries in the wider community. Areas are conceived of as the property of specific groups, in a sense *home ground*. Barnor Hesse[8] explores patterns of habitation, spatial movement and racial violence and Sue Glyptis shows the ways that practices of leisure are historically woven with concepts of *public/private*—and identifies the class and gender relations and processes of urbanisation informing these practices.[9]

Others' Time/Other Time

Karen Davies argues that women's relationship to time is gendered in two ways: *"The dominant temporal consciousness has partly risen out of male interests . . . women's specific subordinate position in society influences how their time may be used and what power they may use in these negotiations."*[10] Drawing on Davies, Barbara Adam argues, *"The working times of women as wives and mothers, both in and out of employment, cannot be placed in a meaningful way within perspectives that separate work from leisure, public from private time, subjective from objective time, task from clock time."*[11] Davies uses the concept of *others' time* to explore the regulation of women's household responsibilities, in particular the time demands of young children. However, this concept is also highly applicable to women's paid work experiences. Struggles around paid work involve selling time to others in a context where there is little power to negotiate over price or place of work. See, for example, Stanlie James's and Abena Busia's[12] discussion of black women's positioning in a peripheral labour force. The concept of *others' time* may be useful in exploring the ways in which women are both spatially positioned and create space for themselves out of *others' space*.

As a result of the hierarchy of values associated with time (see chapter 3, "Spatial/Temporal Concepts Underpinning the Research") the time of paid work is often considered to have more social value than the time of domestic labour. This affects the way that women feel about and conceptualise times of unemployment for example. It also devalues the time they spend in domestic labour. Such factors also impact on the way women feel about leisure and their right to leisure (see below, "Women as Centres of Action"). Eileen Green argues,[13] *"As studies of women's leisure continue to show, time synchronisation and time fragmentation dominate most women's lives . . . which has led them to taking 'snatched' spaces for leisure and enjoyment."* Rosemary Deem[14] stresses the importance of temporal practices to leisure. Women's different personal timetables, the quality of time and simultaneity of activities all impact on leisure experiences.

The corollary of the concept of *others' time/space* is *time/space for self*. The low status of this concept within the social sciences is evidenced in various ways. For example, Jo Van Every[15] when writing about the difficulties of conceptualising time spent doing housework asks, *"If washing is another's labour, what is washing yourself?"* Deem[16] writes of the difficulties for women of achieving leisure time as *time for self* as it is so contingent on the actions of others.

Research Questions—Spatial/Temporal Practices

Critical questions arise from the above research. What forms of segregation/separation in the spheres of home, paid work and leisure do women experience and how does this relate to their studies in higher education? How do public/private practices shape their studies? For example, are some places and times more acceptable as places and times to study? How do other people's space and time, and space and time for other things, impact on studies? The overarching research question is, *"What are the dominant spatial/temporal practices shaping women's everyday experiences whilst studying?"* This question is addressed through the data in chapter 7.

Gender and Spatial/Temporal Representations

Space as Process, not Object

Davina Cooper[17] argues that an adequate understanding of space is achieved by recognising that identified spaces (for example, public space) are *"process not object."* Disadvantaged people are associated with space at the margins[18] and *"social routes through space are shaped and mediated by the organising principles of gender, class and race."*[19] Linda McDowell[20] discusses this in relation to the significance of spatial representations of workplace and gender: *"The location and physical construction of the workplace . . . also affects as well as reflects the social construction of work and workers and the relations of power, control and dominance that structure relations between them."* She discusses the gender coding of particular jobs, and argues that paradoxes related to *"what is considered reasonable behaviour . . . inscribes the division of labour."* Graham Mowl and John Towner[21] raise parallel issues in considering issues of gender, leisure and place: *"The feelings different places engender determine whether what we experience is experienced as leisure or as something completely different."* For example, male dominated planning, physical design and structure, and fear of male or racial violence shape the development of leisure space.

Codings of Social Division

In relation to gendered coding, a number of feminist writers have explored the issue of domestic design and architecture and the powerful symbolic meanings attached to the home and the artefacts it contains. For example Renee Hirschon[22] discusses the experience of low-income households in Yerania, a poor district of Athens. Household space is very limited, nevertheless it is considered essential that each household have at least twelve chairs, a double bed and large dining table. These items are both functional and symbolic. They relate to values and meanings associated with the granting of hospitality to those from outside the home, and harmony between those inside the home.

Marion Roberts[23] has shown the way specific developments in the design of public housing in the U.K. during the 1940s and 1950s reflected particular expectations of gender roles: *"Whilst housing standards may seem to be a political issue connected with class, they also are intimately connected to perceptions of gender difference: in this period, to the notion of the housewife's role in the family."* Ruth Madigan and Moira Monroe argue that assumptions about the use of domestic space and the sharing of labour are embodied in housing design.[24] This is imbued with notions of masculinity and femininity. The home is both a physical space and a *"value laden symbol"*[25] which is integrally connected to the wider sphere, *"His . . . stirring career away from home renders home to him so necessary a place of repose where he may take off his armour, relax his strained attitude and surrender himself to perfect rest . . . It is not her retreat, but her battleground . . . her arena, her boundary, her sphere; to a man it is life in repose; to a woman the house is life militant."*[26]

Home reflects the current and former spatial/temporal ordering of production and reproduction. These are represented in the design of houses and in the ways that certain spaces in the house are prescribed. In the home power can be consolidated through the use of territorial tactics. Michel Foucault,[27] drawing on the work of Philippe Aries, has shown the ways in which the design of houses related historically to certain specified behavioural expectations and provided a source of social regulation through the period of industrialisation: *"Then gradually space becomes specified and functional . . . The working class family is to be fixed, by assigning it a living space with a room that serves as a living space, with a room that serves as a kitchen and dining room, a room for the parents which is the place of procreation and a room for the children, one prescribes a form of morality for the family."* There is a whole history yet to be written in relation to the demarcation of and struggles over space within the household, or, as he puts it, the spatial *"tactics of the habitat."* Conceptualisations of public/private and the gender coding of spaces in the home shape experience and interaction in both home and community. The home symbolically, functionally and technically connects with the wider community. For example, the association of maleness with the exterior rather than interior house,[28] choices about location of home being shaped by decisions about work and family[29] and the technical scientific management of home.[30]

Codes of leisure are associated with employment and with free time (see chapter 4, "Gender and Spatial/Temporal Practices"). Stanley Parker[31] develops a model of leisure that is inextricably woven with paid employment. Leisure is theorised as an extension, opposition or neutral in relation to paid work. Paul Corrigan[32] points out the inadequacy of such a conceptualisation. If leisure is free time then why is unemployment not leisure time? Unemployment is not associated with leisure but with poverty and humiliation: *"It is only possible to understand any aspect of leisure as a result of the major material themes contained within work, domestic labour or school."* Betsy Wearing[33] discusses the gender implications of the concept of leisure as *relative freedom* and as integrally tied to work. Formal leisure is in fact not of women's own choosing. Leisure experiences relate to perception and context—for example eating may be male leisure and simultaneously women's work. Leisure therefore may be both a source of regulation and an entitlement. The ambiguity of leisure for women is summed up by Jennifer Hargreaves[34] who discusses the pleasure, sensuality and fulfilment of exercise and sports for women and at the same time the sexploitation and erotisisation of such activities.

Codings and representations of "race" and nation are also spatial/temporal. Paul Gilroy[35] argues that diversity, change and mobility over time and place undermines nationalist positions, which tend to focus on static concepts and racial identities fixed in time: *"The idea of a common invariant racial identity capable of linking divergent black experience across different spaces and times has been fatally undermined."* He points out that *"simplistic black/white binaries"* are reinforced through nationalist positions.[36] *"However, there is no suggestion that a political language based on racial identification should be abandoned. It may be insufficient but it remains necessary in a world where racisms continue to proliferate and flourish."*[37] Codings of racial division signify certain spaces as the territory of certain groups defined along racial lines.[38]

Research Questions—Spatial/Temporal Representations

The following questions arise. What are the dominant symbolic meanings and codings associated with the different spheres women occupy within paid work, home and leisure/community? How do these relate to women's attempts to integrate higher education into their lives? What lines of power and social divisions are represented in these spaces? What areas of *emotional dissonance* arise for women students in relation to these and their attempts to study? The overarching research question is, *'What are the dominant spatial/temporal representations shaping women's everyday experiences whilst studying?'* This question is addressed through the data in chapter 8.

Women as Centres of Action

Women's Feelings and Conceptualisations

How do women conceptualise home? Whereas women conceptualise home in terms of identity, men are more likely to conceptualise it in terms of ownership: *"Women feel that the meaning of home involves a sense of belonging, a part of 'me,' a refuge, a place to care about . . . sharing of emotions, warmth and security, a core of my existence . . . Men are more likely to relate the meaning of the home to its being a place where things belong to me, feels like its mine, architectural design, a room, a street where I spent my childhood, where one's parents live . . . same place for a long time, leisure."*[39] The research on violence against women in the home has amply demonstrated that the meaning of home does not necessarily relate to what goes on there.

What does leisure space mean for women? *"According to women interviewed . . . leisure is variously: a state of mind, a feeling of personal well being and satisfaction, time free from constraints, anything you choose to do, time when you are relaxed."*[40] Liz Stanley[41] argues, *"Many possible conjunctions of work and leisure are possible and these multiple conjunctions need to be at the heart of our understanding of what it is to be a woman."* To conceive of leisure, women need to conceive of their own time, rather than others' time. However conceptualisations of leisure are shaped by conceptualisations of time for others.[42]

In relation to the spaces and times of paid work, Arlie Hochschild[43] asks whether changing patterns of women's participation in work, and the struggle for time, have changed some women's conceptualisations of work. Work may be increasingly conceptualised as a *"haven from the family," "for some people work may be becoming more like family, and family life more like work."* Here we can see that the process of interpreting space relates closely to changing patterns of social participation.[44] Karen Davies[45] explores the impact of paid work in women's conceptualisations of home. For the unemployed women she interviewed, *"life means less without paid work,"* but women express a *"duality of feelings about home and work"* because they also value more time for themselves and their families. She argues that women's greater participation in paid work relates to the increased value they attached to it.

Women's Actions for Change

"In the house control can be asserted by claiming a room, but it can also be manifested in far more subtle ways . . . the casual out-of-place position of the particularly comfortable chair where the husband reads his newspaper . . . the male's working desk in the living area confines the female to activities in the bedroom or kitchen."[46] Women are able to subtly assert their presence despite rigid segregation. For example, the women of Doshman Ziari in Iran, despite their apparently *"complete exclusion from decision making spaces"*[47] subtly transfer information to men and thereby participate in decision making. Women interact during the day time, and although they are removed from public spheres at dusk[48] they communicate the shared knowledge of women, developed during the day, to individual men, who may take this into the exclusively male domain.[49] Women's impact on housing is also evident. Elizabeth Coit (1938) argued for a radical breakdown of the specific functions of domestic spaces, for example, that the *"nomenclature of rooms be changed or abandoned because the name of each room had potentially restrictive connotations that would at least mentally stifle a more flexible use of spaces. She called for the legitimisation of the kitchen as a space for dining, for study and socialising in bedrooms and for the accommodations of such real life needs as a bedroom or study in the living room."*[50]

In terms of the spheres of paid work, both Patricia Hewitt[51] and Karen Davies[52] discuss the impact of women's greater participation in the labour market on dominant *spatial/temporal practices*. Davies shows that women's choices about work involve resistance to *male time* through, for example, rejecting wage labour, prioritising time over money and prioritising part-time work.[53]

Leisure is a site of change for women.[54] It contains the possibility for women to become themselves, to develop *"alternative self definitions and identities."*[55] Voluntary work or *"serious leisure"* may be a *"site for personal fulfilment identity enhancement and self expression."* Leisure engagement may be conceptualised as a political act[56] being *purposive* and *"a means of taking control and finding meaning"*[57] involving negotiation and collaboration. A good example of this is from Eileen Green's research.[58] She discusses the significance of friendship as a leisure choice for women. Friendship, she argues, is both process and play. It is a *"collaborative tool for exploring the world,"*[59] and *"a means of mediating public and private subjectivities."*[60] Erica Wimbush points out that women may also resist scrutiny through leisure practices.[61] Space and time are critical to negotiating leisure. Susan Shaw[62] points out that *"the political nature of leisure [is] an aspect of leisure which is not always recognised or understood"* and stresses the need to ground and analyse how resistance (collective or individual) manifests itself in leisure practices. How, for example, are *"specific types of oppression or constraint . . . being challenged or resisted?"*[63] What is the role of *intention* in this process? What types of *outcome* arise and how can we understand these?

The issues raised in feminist leisure theory are very close to those concerned with creating space and time for study in higher education (see chapter 8). There is a growing body of feminist research which focuses on the experiences, feelings and actions of women in reshaping and restructuring their social worlds. Women are involved in conceptualising and coding space, struggling for multifunctional and safe space, weaving the multiple time demands of others and re-evaluating the social meaning of space and time for themselves. In all these ways individual women and groups of women are vivid centres of action in transforming the landscapes they experience.

Research Questions—Women as Centres of Action

Critical questions arise from the above in relation to women's attempts to study in higher education. Struggles for space and time in the spheres of home, paid work, leisure/community are involved. How do women conceptualise higher education in relation to these other spheres? Do they consider that their higher education is of benefit to the home and household or is it conceptualised as a haven from home and an escape from domesticity? Is higher education conceptualised as advancing opportunities of paid work or as an escape from the drudgery of paid work? Is it perceived as a form of leisure for themselves or as intruding on leisure opportunities? Fundamentally, is higher education conceptualised as their right, or as a burden on themselves and others? In addition, what activities do women engage in to create the space and time for studies? What re-ordering of activities and negotiations go on in the different spheres of home, paid work and leisure/community? How do women construct physical places and times for their studies? How do they negotiate and snatch space and time for studies? In what ways do they collaborate with others including family and friends? Where do they find resistance? How do they construct a rationale to justify studying and in what ways does this involve reshaping their sense of self and others? What rhythms and rituals do they develop to support their studying? The overarching research question is, *"What actions (physical, mental and emotional) do women of differing social position take in order to integrate higher education into their lives?"* This question is addressed through the data in chapter 10.

Conclusion

I have drawn out key research questions that will make it easier to address what is involved when women seek to bring higher education into their lives. Central to this, I have discussed the criticality of including spatial/temporal concepts in research with women. It is also essential that questions be asked about the ways that space and time have actually shaped the construction, process and outcomes of the research itself. The following critical questions therefore arise: *"How have space and time shaped the research?"* and *"What are the implications of the researcher's action for the findings?"* These questions are discussed respectively in chapters 6 and 11. In the following chapter I review literature concerned with women and higher education, building on ideas and questions from this chapter to develop further questions in relation to the institution of higher education itself.

Notes

1. Daphne Spain, *Gendered Spaces* (Chapel Hill: The University of North Carolina Press, 1992).
2. Eileen Green, Sandra Hebron and Diana Woodward, "Women, Leisure and Social Control," in *Women, Violence and Social Control*, edited by Jalna Hanmer and Mary Maynard (London: Macmillan, 1987), 128–29.

3. Tamara Dragadze, "The Sexual Division of Domestic Space among Two Soviet Minorities: The Georgians and the Tadjiks," in *Women and Space: Ground Rules and Social Maps*, edited by Shirley Ardener (Oxford: Berg, 1993), 159–63. She explores the different experiences of Georgian and Tadjik communities. Although Tadjik women experience a more explicit spatial segregation, for example, having to retire if a male visitor comes to the house (161), the consequence of this is that men are far more likely to be involved in domestic activities, for example cooking, making tea and sweeping in the company of other men (162). This is in order to avoid women's presence. On the other hand, Georgian women experience far less spatial confinement (159), being allowed to work everywhere both in and out of the house. Nevertheless, there are much stronger divisions of domestic labour (163). She accounts for these differences in spatial use by relating them to differences in kinship structure and the regulation of marriage.

4. Peter Williams, "Constituting Class and Gender: A Social History of the Home 1700–1901," in *Class and Space: The Making of Urban Society*, edited by Nigel Thrift and Peter Williams (London: Routledge and Keegan Paul, 1987), 157.

5. Garry Whannel, "Electronic Manipulation of Time and Space in Television Sport," in *Leisure, Time and Space: Meaning and Values in People's Lives*, edited by Sheila Scraton (Eastbourne, U.K.: Leisure Studies Publications, 1998), 9.

6. Patricia Hill Collins, *Black Feminist Thought: Knowledge, Consciousness and the Politics of Empowerment* (London: Unwin Hyman, 1990).

7. Eva Gamarnikow, David Morgan, June Purvis and Daphne Taylorson, eds., *The Public and the Private* (England: Gower, 1983).

8. Barnor Hesse, "Racism and Spacism in Britain," in *Tackling Racial Attacks*, edited by Peter Francis and Roger Matthews (Leicester: C. S. P. O. publishers, 1993).

9. Sue Glyptis, *Leisure and Unemployment* (Milton Keynes: Open University Press, 1989), 5.

10. Karen Davies, *Women and Time: The Weaving of the Strands of Everyday Life* (Aldershot: Avebury, 1990), 9.

11. Barbara Adam, *Timewatch: The Social Analysis of Time* (Cambridge: Polity Press, 1995), 95.

12. Stanlie M. James and Abena P. A. Busia, eds., *Theorising Black Feminism: The Visionary Pragmatism of Black Women* (New York: Routledge, 1993), 19.

13. Eileen Green, "Women Doing Friendship: An Analysis of Women's Leisure as a Site of Identity Construction, Empowerment and Resistance," *Leisure Studies* 17, (1998): 171–85.

14. Rosemary Deem, "Feminism and Leisure Studies: Opening up New Directions," in Relative *Freedoms: Women and Leisure*, edited by Erica Wimbush and Margaret Talbot (Milton Keynes: Open University Press, 1988), 11.

15. Jo Van Every, "Understanding Gendered Inequality: Reconceptualising Housework," *Women's Studies International Forum* 20, no. 3 (1997), 411–20.

16. Deem, "Feminism, Leisure."

17. Davina Cooper, "Regard between Strangers: Diversity, Equality and the Reconstruction of Public Space," *Critical Social Policy* 18, no. 4 (November 1998), 466.

18. Cooper, "Public Space," 468.

19. Cooper, "Public Space," 473.
20. Linda McDowell, *Capital Culture. Gender at Work in the City* (Oxford: Blackwell, 1997), 12.
21. Graham Mowl and John Towner, "Women, Gender, Leisure and Place: Towards a More Humanistic Geography of Women's Leisure," *Leisure Studies* 14, no. 2 (April 1995): 102–16, 104.
22. Renee Hirschon, "Essential Objects and the Sacred: Interior and Exterior Space in an Urban Greek Locality," in *Women and Space: Ground Rules and Social Maps*, edited by Shirley Ardener (Oxford: Berg, 1993).
23. Marion Roberts, *Living in a Man-Made World: Gender Assumptions in Modern Housing Design* (London: Routledge, 1991), 78.
24. Ruth Madigan and Moira Monroe, "Gender, House, and 'Home': Social Meanings and Domestic Architecture in Britain," in *Women and Social Policy: A Reader*, edited by Clare Ungerson and Mary Kember (London: Macmillan, 1997), 205.
25. Susan Saegart and Gary Winkel, "The Home: A Critical Problem for Changing Sex Roles," in *New Space for Women*, edited by Gerda R. Wekerle, Rebecca Peterson and David Morley (Boulder, CO: Westview Press, 1980), 41.
26. Cynthia Rocke, Susanne Torre and Gwendolyn Wright, "The Appropriation of the House: Changes in House Design and Concepts of Domesticity," in *New Space for Women*, edited by Gerda R. Wekerle, Rebecca Peterson and David Morley (Boulder, CO: Westview Press, 1980), 84, quote from Appleton's Journal (1870).
27. Michel Foucault, "Truth and Power," in *Power/Knowledge: Selected Interviews and Other Writings 1972–1977 by Michel Foucault*, edited by Colin Gordon (New York: Pantheon Books, 1980), 149.
28. Madigan and Monroe, "Gender, House."
29. Gerda R. Wekerle, Rebecca Peterson and David Morley, *New Space for Women* (Boulder, CO: Westview Press, 1980).
30. Williams, "Class, Gender."
31. Stanley Parker, "Towards a Theory of Leisure and Work," in *Sociology of Leisure: A Reader*, edited by Charles Critcher, Pete Bramham and Alan Tomlinson (London: E. and F. N. Spon, 1995), 29.
32. Paul Corrigan, "The Trouble with Being Unemployed Is That You Never Get a Day Off," in *Freedom and Constraint: The Paradoxes of Leisure: Ten Years of the L.S.A*, edited by Fred Coalter (London: Routledge, 1982/1989), 193.
33. Betsy Wearing, *Leisure and Feminist Theory* (London: Sage, 1998), xii, 25–37.
34. Jennifer Hargreaves, *Sporting Females: Critical Issues in the History and Sociology of Women's Sports* (London: Routledge, 1994), 161–66.
35. Paul Gilroy, *Small Acts: Thoughts on the Politics of Black Culture* (London: Serpent's Tail, 1993), 2.
36. Gilroy, *Small Acts*, 31.
37. Gilroy, *Small Acts*, 14.
38. See Chapter 6 for a discussion of concepts of "racial" difference in the research.
39. Saegart and Winkel, "The Home," 47.
40. Mowl and Towner, "Gender, Leisure," 111.

41. Liz Stanley, "Historical Sources for Studying Work and Leisure in Women's Lives," in *Relative Freedoms: Women and Leisure*, edited by Erica Wimbush and Margaret Talbot (Milton Keynes: Open University Press, 1988), 19.

42. Sheila Scraton and Beccy Watson, "Gendered Cities: Women and Public Leisure in the 'Post Modern City,'" *Leisure Studies* 17, no. 2 (April 1998): 123–37.

43. Arlie R. Hochschild, "The Emotional Geography of Work and Family Life," in *Gender Relations in Public and Private: New Research Perspectives*, edited by Lydia Morris and E. Stina Lyon (London: Macmillan, 1996), 28.

44. Eileen Green and Sandra Hebron, cite Westwood "Leisure and Male Partners," in Relative *Freedom: Women and Leisure*, edited by Erica Wimbush and Margaret Talbot (Milton Keynes: Open University Press, 1988).

45. Davies, *Women, Time*, 138.

46. Rocke, Torre and Wright, "Appropriation, House," 94–95.

47. Susan Wright, "Place and Face: Of Women in Doshman Ziari, Iran," in *Women and Space: Ground Rules and Social Maps*, edited by Shirley Ardener (Oxford: Berg, 1993), 150.

48. Wright, "Place, Face," 152.

49. Wright, "Place, Face," 150.

50. Rocke, Torre and Wright, "Appropriation, House," 93–94, discuss the work of Elizabeth Coite.

51. Patricia Hewitt, *About Time. The Revolution in Work and Family Life* (I. P. P. R., London: Rivers Oram Press, 1993).

52. Davies, *Women, Time*.

53. Davies, *Women, Time*, 204–6.

54. Wearing, *Leisure, Feminist*, 43.

55. Wearing, *Leisure, Feminist*, 46.

56. Josephine Burden, "Leisure, Change and Social Capital: Making the Personal Political" (paper presented at Leisure Studies Association International Conference "The Big Ghetto" [July 1998]), 1.

57. Burden, "Leisure, Change," 10.

58. Green, "Women, Friendship," 176.

59. Green, "Women, Friendship," 179.

60. Green, "Women, Friendship," 180.

61. Erica Wimbush, "Mothers Meeting," in *Relative Freedoms: Women and Leisure*, edited by Erica Wimbush and Margaret Talbot (Milton Keynes: Open University Press, 1988), 65.

62. Susan Shaw, "Conceptualising Resistance: Women's Leisure as Political Practice," *Journal of Leisure Research* 33, no. 2 (2001): 186–201, 198.

63. Shaw, "Conceptualising Resistance," 199.

5

The Spatial and Temporal Relations of Higher Education

"There'd be a couple of teachers that would look at me stupid. 'What are you doing here? You're wasting my time!'"

Respondent Lorna, chapter 9

"Still worried about questionnaire . . . can't give it out till [supervisor's] commented. Was really glad [supervisor] thought feminist theory paper OK. [Supervisor] boosted my confidence . . . bit on tenterhooks till I hear from [supervisor]."

Researcher Log, chapter 11

Introduction

HAVING IDENTIFIED KEY QUESTIONS arising from research into women's experiences in the spheres of home, paid work and leisure/community, I now discuss research directly concerned with higher education and gender. The aim is to review research and identify further research questions drawing on concepts discussed in chapter 3. The chapter is in three parts considering: spatial/temporal practices and higher education; spatial/ temporal representations and higher education; and women students' perceptions and experiences.

Higher education has become increasingly formalised and regulated in advanced industrialised societies. This is a result of its relationship to the state and wider labour market. Political perspectives on higher education reflect differing beliefs about social and economic development as a whole. Aiden Foster Carter,[1] for example, discusses the conflicting views of modernisation and dependency theorists in relation to higher education in the southern hemisphere. Whereas modernisation theorists emphasise the role of higher education in creating human capital for the economy, nation building and developing a *modern consciousness*, dependency theorists emphasise the creation of academic elites and the pressures through higher education to cultural conformity.

Research into women and higher education has covered wide ground. There has been considerable focus on the particular difficulties faced by mature students in relation to balancing finance and the demands of home, paid work and higher education.[2] There has been less focus on the circumstances of younger women students. Within women's studies in particular, the effectiveness of teaching and learning strategies and the relevance of the dominant pedagogies and epistemologies of higher education to women's own knowledge and experience have been discussed.[3] In addition researchers have examined local, national and international developments in education policy in relation to institutional practice.[4] Critical questions have been raised regarding the impact of higher education for women. Does higher education perpetuate and reinforce women's social position? How do women transform their position through the higher education process? Several writers have argued for further research into the ways in which women's everyday lives connect with higher education. It is felt that this would give more insight into effective learning and teaching methods and methods of recruitment.[5]

Feminist perspectives on education have been dichotomous. On the one hand, some writers identify higher education as a sphere that perpetuates gendered inequalities and reproduces the social relations of wider society.[6] Certain academic and paid employment routes are prescribed in preparation for gendered horizontal and vertical segregation in the labour market. On the other hand, research with women students has shown the ways they themselves identify higher education as liberating them from domesticity.[7] Ann White[8] argues that education theorists, with exceptions, *"have either tended to emphasise social structure and ignore agency, or in ethnographic studies, if social structure is addressed either no attempt is made to trace the links which exist between structure and action or human agency is dissolved into the notion of social structure."* Higher education can be conceptualised as a site of production and consumption in relation to knowledge and skills[9] and as a site of the reproduction of human and cultural capital.[10] Students in higher education are increasingly conceptualised as consumers, yet their relationship to the institution is ambiguous as they actively engage with the production of knowledge.

Feminists characterise higher education as a site of struggle where values and meanings are contested.[11]

Spatial/Temporal Practices and Higher Education

Barbara Adam[12] argues, *"Even the most cursory examination of the way education is organised in Western style societies shows that everything is timed. The activities and the interactions of all its participants are orchestrated to a symphony of buzzers, bells, timetable schedules and deadlines."* The organisation of higher education echoes patterns of economic production for both pragmatic and strategic reasons. Workers (and students) engaged in higher education are necessarily subject to the wider rationalities of work and society. Developments in law and policy affecting the wider labour market affect them also. In addition, higher education provides a strategic role for the state and capital in providing a skilled, flexible workforce. The regulation of higher education involves political decisions about which social actors will be involved in policy development and management. Education reforms in the U.K.[13] have increased the involvement of private business in the management of higher education.

Spatial practices in higher education also share characteristics with economic production. For instance, the large educational institution is the centre of the educational process, marked by historical patterns of geographic and architectural segregation. In most countries of the world access to higher education for women has been segregated: *"Colleges were closed to women until the late nineteenth century because physicians believed that school attendance endangered women's health and jeopardised their ability to raise children."*[14] Segregation in higher education also operates along the axes of class, "race," age and dis/ability.[15] Daphne Spain argues that the history of declining educational sexual segregation in the North relates to the increasing integration of the work force. Nevertheless patterns of sexual and other segregation persist in relation to access, curriculum choice and staffing.

There have been major shifts within higher education in relation to the deregulation of the labour market and globalisation. Crescy Cannan discusses the ways in which *enterprise culture* has been promoted within higher education and argues that this relates to capital's need for a more tightly regulated labour force in the context of an increasingly de-regulated labour market. These developments have affected the spheres of economic production, social welfare and education. She discusses the developing links between the institutions of education and labour, tighter monitoring of those links, and increased pressure to break down the educational process into explicit units of regulation. She argues that the concept of a *student* in higher education is now embedded with ideas about *"the enterprising self,"* drawn from the political ideology of neo-liberalism, which emphasises *"autonomy, responsibility, initiative, self-reliance, independence and willingness to take risks, to 'go for it', see opportunities and take responsibility for one's own actions."*[16] The U.K. Government, under New Labour,[17] has focused on the themes of widening access, lifelong learning, students as consumers and links with the labour market. Such initiatives have the capacity to transform the dominant meanings associated with the education of adults and tie these more closely to concepts of work and the work ethic. At the same time parts of the education system are being explicitly linked to welfare eligibility both in the U.S. and the U.K. under the new *workfare* systems. In addition, the reduction of state financial support to higher education students has led to a heavier reliance by them on paid work whilst studying.

Lena Dominelli and Ankie Hoogvelt draw direct connections between developments in higher education (in particular social work education) and the globalisation of capital. The demands for tighter regulation and product evaluation relate to *"A historical trend towards forms of production organisation in which capital no longer needs to pay for the reproduction of labour power."*[18] The role of higher education in the reproduction of labour is accentuated in certain ways, for example the emphasis on transferable skills, lifelong learning, and flexibility. The state is actively involved in promoting such developments. In the early 1980s the Welfare State was *"actively drawn into the market place." "Business management techniques were brought into welfare services, particularly education and health."*[19] Although Dominelli and Hoogvelt are particularly concerned with developments in the training and education of social workers, the processes they identify are reflected in all spheres of higher education, where there is an increased emphasis on *"performance targets, input and output measures, costings, value for money . . . roles and responsibilities identified . . . monitoring and reporting mechanisms."*[20] They describe these developments as a form of *Taylorisation*, concerned with the tighter regulation of social welfare. The focus is on temporal regulation.

A raft of changes has therefore reshaped patterns of *spatial/temporal practice* in higher education. These include:

- Widened access without associated increases in physical space or staffing.
- The restructuring of the curriculum through modularization, credit accumulation and semesterisation. This may facilitate flexibility and student choice, but at the expense of course coherence. It leads to unpredictability and increased pressures on space and time. There is also the potential for fast tracking of students (for example two-year degree programmes).

- The cuts in financial support to students have created additional time pressures for students who must find an alternative income.
- The introduction of business management techniques has generated an additional administrative load for both clerical and academic staff.
- Technological developments, which potentially facilitate student centred learning (distance learning, home-based learning) have had effects regarding student access to resources, student/tutor face-to-face contact and the staffing and resourcing of higher education as a whole.

One aspect of the above developments is the growing spatial distance between tutor and student in higher education. Increased student numbers, technological developments and the growth of distance-based and student-led practices has led, in some cases, to less contact between tutors and individual students. This is not an inevitable consequence of such developments and more support in the area of study skills has been initiated in many institutions.

Clearly the *spatial/temporal practices* of higher education are being transformed in a number of ways. The *clock-led* rationalities still dominate, but tighter regulation of the educational process, increased emphasis on specific labour market needs, decreased student finance and technological developments are all having a major impact. The dominance of the world of work is evident in three ways. Firstly, both staff and students are subject to the wider clock based rationalities of work. Secondly, the changing world of work shapes the educational experience through the restructuring of the meaning of higher education, course content and methods of teaching and learning. Thirdly, student reliance on low paid work is beginning to be conceptualised as an educational resource rather than as a burden for those students involved.

Research Questions—The Spatial/Temporal Practices of Higher Education

Several critical questions arise from the above literature. How do the *clock-led* rationalities of higher education impact on women students, for example, in relation to assessment, progression and semesterisation? How do changes in student numbers shape interaction, for example, access to tutors and learning support? What has been the impact of enterprise culture and the links with employment and business? How has the reduction of state financial support to students impacted on women's experiences of higher education? The underpinning question is, "*What are the dominant spatial/temporal practices of higher education shaping women's experiences?*" This question is addressed through the data in chapter 9.

Spatial/Temporal Representations and Higher Education

The places of higher education (physical and academic) have been produced over time and are embedded with meanings from past times. Above I discussed the ways that current educational practices are being transformed. The concept of representations gives visibility to the outcome of past transformations in higher education as they are represented in the current curriculum and learning environment. This section is in two parts and discusses: the curriculum and academic disciplines; the learning environment and teaching methods.

The Curriculum and Academic Disciplines

Kim Thomas[21] explores the experiences of male and female students and the meanings they attach to particular aspects of the curriculum. Her research focuses on the polarities of *masculinity/femininity* and of *science/arts*. She demonstrates the ways in which these polarities have historically shaped the current curriculum and in turn shape the choices of students. She argues that although active discrimination against women in higher education has reduced, it is harder for women to succeed in particular disciplines and at particular levels because of the history of the curriculum and current perceptions of this. Academic disciplines emerge historically and are shaped by struggles over meaning. Feminists have pointed to the limitations of the existing disciplines as they represent past masculine, imperialist and class-based ideas about the social world. For example in the disciplines of economics and home economics, household relations are separated from the economics of the public sphere. Feminists have argued for new academic disciplines, for example Women's Studies, but also stress the need for multidisciplinarity as a means to developing more realistic worldviews.[22] Although there has been an increased representation of women in higher education, there is still a striking imbalance in relation to the spheres they occupy and the status accorded to those spheres. They lack representation in the natural sciences and engineering and are over-represented in health, social and behavioural sciences.[23] *"Education is the critical line between public and private status for women,"*[24] and although sex segregation has decreased, *"status will only improve if access to knowledge is achieved."*[25] A range of social divisions are represented in the curriculum, including ethnocentrism and ageist, sexist and class-based constructs which are *"frequently normalised in educational discursive practices."*[26] The growth of specific routes through higher education in relation to gender, black studies and workers' education has created specific spaces along lines of social division. It has been argued that these spaces operate as sites of change within the institution and have positive effects for higher education as a whole. However, the potential danger of such developments is pointed out by C. L. R. James: "Now to talk to me about black studies as if it's something that concerned black people is an utter denial. This is the history of western civilisation. I can't see it otherwise. This is the history that black people and white people and all serious students of modern history and the history of the world have to know. To say it's some kind of ethnic problem is a lot of nonsense."[27]

Learning Environment and Teaching Methods

Particular teaching methods in the academy have historically disfavoured women and working class students.[28] These are didactic, non-collaborative and require conformity. Higher education represents a number of pedagogic traditions. That which dominates, particularly in the older universities, is characterised by the concept of the lecturer as a source of expert knowledge, a hierarchy of occupations where original research has precedence over teaching, and a model of learning based on individual achievement, competition and adversarial methods of knowledge creation.[29] There are clearly other pedagogic traditions within higher education, which have often developed from the smaller colleges, the merged further/higher education institutions and the more imaginative approaches in workers' and women's education. Nevertheless, these can be undermined by the more dominant tradition. The learning environment represents these ideas in its physical make-up. For example, the formal lecture theatre facilitates one way communication rather than the sharing of ideas. The library is a place of silence rather than a place for collaborative work. Despite these educational practices being transformed, the physical buildings and layout of rooms represent past orthodoxies and have educational effects.

Sue Jackson[30] considers these issues in her paper "Safe Spaces: Women's Choices and Constraints in the Gendered University." Through interviews with women involved in women's studies programmes she explores women's experiences in corridors, in individual interaction with tutors and in the classroom. She draws attention to the denigration of some women's experiences as non-academic and to the dominant masculine culture within the institution, which affects all teaching. She argues that space in the University is gendered in four ways, physically, geographically, emotionally and intellectually. Space in the University is also gendered in relation to the experience of women academics. Researchers highlight emotional and intellectual segregation, women's academic isolation[31] and balance of power issues between staff and students, including for example, the control that may be exercised by men students.[32]

Research Questions—The Spatial/Temporal Representations of Higher Education

Critical questions arise from the above literature. What polarities exist in the curriculum and curriculum choice in relation to social divisions? What teaching and learning strategies are applied and how are these experienced? What settings, spaces and artefacts are used in teaching and how are these perceived and engaged with? The underpinning question is, *"What are the dominant spatial/temporal representations of higher education shaping women's experiences?"* This question is addressed through the data in chapter 9.

Women Students' Experiences and Perceptions

Key critical questions arise as to how women conceptualise time spent within higher education. In paid work women sell their time to others. At home, their time may be given and/or coerced. What is educational time? How do women rationalise this? Is it conceptualised as time for others? Is it an opportunity to increase their bargaining power over the value of their own time in the labour market or as time for themselves? How do they create space and time for studies? Experiences related to "race," age and ability are equally relevant: *"perceptions of self are formed and shaped by structural factors . . . which in turn give rise to perceptions of structure."*[33] Beverley Skeggs,[34] for example, discusses the way class operates not just as a context of particular behaviour but operates as a *"structure of feeling"* shaping decisions. Women are involved in reconceptualising their social, spatial and temporal world in order to create space and time for studies. Three areas of research are now discussed: deciding to enter higher education; perceptions of the curriculum and learning environment; and creating space and time to study.

Deciding to Enter Higher Education

Ann White argues that academic analysis of decisions to enter higher education have been dominated by the psychology of motivation studies. This *"tends to emphasise the individual rather than the group and not recognise the social action of returning to education."*[35] A social model of motivation is adopted by Heidi Mirza who identifies a *"collective educational urgency among young black women to enter colleges of further and higher education."* Although some commentators explain this as attempts to conform, Mirza argues that young black women are *"trying to create spheres of influence that are separate from but engaged with existing structures of oppression . . . Black women strug-*

gle for educational inclusion in order to transform their opportunities and so in the process subvert racist expectations and beliefs."[36] In considering motivation, Gillian Pascall and Roger Cox explore the combination of factors that shape white women's decisions to return to education. They point to a web of family based and economic factors at play including the frustrations of low paid work, family breakdown, the need for autonomy, negative feelings about the housewife role and mixed feelings about mothering. Higher education is conceptualised as *"a starting point of a new process breaking out of domesticity and low paid work."*[37]

Kalwant Bhopal writes of the pressures on young Asian women to settle down and marry just at the stage higher education commences.[38] Uma Narayan writes similarly of the ambiguous response to education from mothers: *"Both our mothers and our mother cultures give us all sorts of contradictory messages . . . for example praising education but fearing the consequences for their daughters."*[39] bell hooks echoes this: *"Many black parents have encouraged children to acquire an education while simultaneously warning us about the dangers of education."*[40] Higher education is here conceptualised as a route out of oppression which may also involve separation from the communities with which women identify and the associated knowledge and understanding generated by those communities.

The significance of higher education as a route to paid employment is stressed by Karen Davies.[41] Working class women she interviewed saw a return to education as *time wasted*, as they did not see it as a route to paid work. Ann White points out that women may rationalise decisions to return to education in terms of *others' time*, *"It would appear that even when women take the decision to return to education, when they become motivated and decide to act for change, that decision, for many women, needs to be rationalised in terms of family and/or work responsibilities rather than in terms of personal needs or desires."*[42]

Perceptions of Curriculum and Learning Environment

Rosalind Edwards[43] discusses women's perceptions of the relatedness of the curriculum to their other experiences. Her research with mature mother students shows that women perceive a hierarchy of values around time spent on certain activities. Experiences of paid work are drawn on by academics and offered by students in the classroom, but experiences of mothering are not. Students feel mothering is devalued and considered inferior and demeaning in academic conversation.[44]

In terms of the learning environment, bell hooks discusses the perceptions of black women students in higher education, *"Many of us have found that to succeed at the very education we had been encouraged to seek, would be most easily accomplished if we separated ourselves from the experience of black folk, the underprivileged experience of the black underclass that was grounding our reality.*"[45] This may result from academic practices that *"encourage academics to treat other people's words and experiences as resources which can be selectively organised to illustrate pre-determined conceptual frameworks."*[46] Black students, working class students and disabled students may be categorised or objectified in narrow ways both by peers and lecturers. Mother students' perceptions of the learning environment may reinforce their perception of the low status of motherhood in higher education[47]; for example, the lack of freely available childcare and the lack of advance timetables so that women can plan childcare well ahead confirm this. In addition, perceptions of the learning environment may be radically changed through experiences of violence against women, either on campus or outside the institution.[48]

Creating Space and Time to Study

In order to create space and time to study, students are involved in reconceptualising and reordering space and time in other spheres of their lives. The focus of research on women and higher education has been primarily on their experiences of the higher education institutions. Some writers have researched the connections between the two spheres of higher education and household.[49] Rosalind Edwards discusses the complex ways in which women separate and connect the spheres of family and higher education, the emotional labour involved and the dynamics in the household. A quarter of her sample split from their partners by the end of their studies. She argues that the family and higher education operate as *"two greedy spheres."* Women are *"time poor"* and their partner's time consumes their attention, *"The possibility that the demands of family and higher education are incompatible in terms of the amount of time each laid claim to was occasionally mentioned by the women themselves, in particular maternal time."*[50] Some male partners saw higher education as *"violating the home as separate and private"* both physically and emotionally. Women had little study space. Some worked in kitchens, some with the television on. Only certain areas of academic study could be discussed with partners without being threatening. Women's attempts to connect the spheres of family and higher education, for example by raising issues in the classroom or at home or by taking family to college or students to their homes, were generally thwarted. Ultimately, *"successful"* students tried to keep these spheres of their lives separate. Edwards argues that higher education research has either explored the dilemmas women present to higher education or the difficulties higher education presents for women students. Neither approach is sufficient because of the complex sexual politics involved. As Madeleine Gromet[51] points out, *"It is important to maintain our sense of this dialectic relation, wherein each milieu, the academic and the domestic, influence the character of the other and not to permit the relation to slide into simplistic one sided causality."* Beyond this, there is a need to explore the ways in which women integrate the multiple demands on their time from paid as well as domestic work, and to explore the ways they mediate, conceptualise and reorder the spheres of higher education, household, paid work and leisure/community.

This account of the literature has paid little attention to women's attempts to transform academia[52] and to the radical changes women have brought about to the practices and representations of higher education. For example, feminists have struggled for appropriate time tabling, adequate childcare, the transformation of the curriculum and collaborative student-centred teaching methods. Sue Jackson[53] writes of women's studies courses in particular as spaces for feminist activism within higher education. As *centres of action*, women have been engaged in reshaping and restructuring higher education. They are involved in reconceptualising and recoding educational space, struggling for multidisciplinarity and re-evaluating the social meaning of education for themselves.

Research Questions—Women Students' Experiences and Perceptions

Critical questions arise from the above literature. What processes give rise to student entry to higher education including those relating to home, paid work and leisure? How do women manage the curriculum and learning environment? How are the spaces of higher education conceptualised and used including halls of residence, the students union and the library? How do women students negotiate space and time for studies with other individuals in the institution including learning support staff, tutors and peers? The underpinning question is, *"What actions (physical, mental and emotional) do women of differing social position take in relation to the institution of higher education in order to create space and time to study?"* This question is addressed through the data in chapter 10.

Conclusion

Kari Delhi argues, *"Knowledge production in universities is a terrain of conflict not just between women and men but among women. Our differences are not just discursive or theoretical; they are material, embodied and political, as we struggle against or conform to ways of knowing and being in the world . . . but also we continue to interrupt, disrupt, subvert as we fail, run away, get sick, feel stressed, have breakdowns.*"[54] The dominant meanings attached to higher education are being transformed. Conceptualisations of work related to capitalist development and the language of consumerism are actively reshaping the *spatial/temporal practices* of higher education. Tighter regulation of the educational process, the use of business management techniques and technology and the values of enterprise culture are now characteristic. Higher education is subject to restructuring in terms of organisation, course content and delivery. Past ideas about gender and other social divisions and associated educational practices are still represented in the curriculum, the learning environment and approaches to teaching. These dominant practices and representations shape women students' experiences of higher education. They are also shaped by the practices and representations involved in the other spheres of their lives. Women mediate, reorder and reconceptualise their social world in order to create space and time for higher education. Below I summarise the key research questions from chapters 4 and 5 and show where they are addressed in the book. In the following chapter I move on to discuss the research methodology.

TABLE 5.1
Summary: Research Questions

Research Questions	*Research Analysis*
What are the dominant spatial/temporal practices shaping women's everyday experiences whilst studying?	Chapter 7: Spatial and Temporal Practices: Frameworks for Action
What are the dominant spatial/temporal representations shaping women's everyday experiences whilst studying?	Chapter 8: Spatial and Temporal Representations: Guidelines for Action
What are the dominant spatial/temporal practices of higher education shaping women's experiences?	Chapter 9: The Framework and Guidelines of Higher Education
What are the dominant spatial/temporal representations of higher education shaping women's experiences?	Chapter 9: The Framework and Guidelines of Higher Education
What actions (physical, mental and emotional) do women of differing social position take in order to integrate higher education into their lives?	Chapter 10: Women as Centres of Action
How have space and time shaped the research?	Chapter 6: Research through the Prism of Space and Time
What are the implications of the researcher's actions for the findings?	Chapter 11: Conclusions: Women Creating Space and Time

Notes

1. Aiden Foster Carter, "The Sociology of Development," in *Sociology: New Directions*, edited by Mike Haralambos (Ormskirk: Causeway, 1985), 181.

2. Gillian Pascall and Roger Cox, "Education and Domesticity," *Gender and Education* 5, no. 1 (1993): 17–35; Rosalind Edwards, *Mature Women Students: Separating or Connecting Family and Education* (London: Taylor and Francis, 1993); Hilary Arksey, Ian Marchant and Cheryl Simmill, *Juggling for a Degree: Mature Students' Experience of University Life* (Unit for Innovation in Higher Education, Lancaster University, 1994).

3. Mary F. Belenkey, Blythe McVicker, Nancy R. Goldburger and J. M. Tarule, *Women's Ways of Knowing: The Development of Self, Voice and Mind* (New York: Basic Books, 1986); Kim Thomas, *Gender and Subject in Higher Education* (Buckingham: The Society for Research into Higher Education and Open University Press, 1990); Himani Bannerji, Linda Carty, Kari Dehli, Susan Heald and Kate McKenna, *Unsettling Relations: The University as a Site of Feminist Struggles* (Toronto: Women's Press, 1991); Joanne De-Groot and Mary Maynard, eds., *Women's Studies in the 1990s: Doing Things Differently* (Hampshire: Macmillan, 1993); Roseanne Benn, Jane Elliot and Pat Whaley, eds., *Educating Rita and Her Sisters: Women and Continuing Education* (Leicester: National Institute of Adult Continuing Education, 1998).

4. Crescy Cannan, "Enterprise Culture, Professional Socialisation and Social Work Education in Britain," *Critical Social Policy* 42 (Winter 1994/1995): 5–18; Lena Dominelli and Ankie Hoogvelt, "Globalisation and the Technocratization of Social Work," *Critical Social Policy* 16, no. 2 (May 1996): 45–62.

5. Veronica McGivney, "Dancing into the Future: Developments in Adult Education," in *Educating Rita and Her Sisters: Women and Continuing Education*, edited by Roseanne Benn, Jane Elliot and Pat Whaley (Leicester: National Institute of Adult Continuing Education, 1998), 16; Viv Anderson and Jean Gardner, "Continuing Education in the Universities: The Old, the New and the Future," in *Educating Rita and Her Sisters: Women and Continuing Education*, edited by Roseanne Benn, Jane Elliot and Pat Whaley (Leicester: National Institute of Adult Continuing Education, 1998), 24.

6. Kim Thomas, *Gender, Subject*.

7. Pascall and Cox, "Education, Domesticity."

8. Ann White, "Perception of Student Needs: A Feminist Approach" (paper presented at "Higher Education Close Up," University of Central Lancashire, 6–8 July 1998), 4.

9. Liz Stanley, "Feminist Praxis and the Academic Mode of Production," in *Feminist Praxis*, edited by Liz Stanley (London: Routledge, 1990).

10. Beverley Skeggs, *Formations of Class and Gender: Becoming Respectable* (London: Sage, 1997).

11. Bannerji, Carty, Delhi et al., *Unsettling Relations*.

12. Barbara Adam, *Timewatch: The Social Analysis of Time* (Cambridge: Polity Press, 1995), 61.

13. See, for example, the *UK Education Reform Act 1988.8*, which weakened the power of academics in educational funding councils and gave more direct power to employers.

14. Daphne Spain, *Gendered Spaces* (Chapel Hill: University of North Carolina Press, 1992), 4.

15. John Bird, *Black Students and Higher Education: Rhetorics and Realities* (Milton Keynes: Open University Press, 1996); Mike Oliver and Colin Barnes, *Disabled People and Social Policy: From Exclusion to Inclusion* (London: Longman, 1998).

16. Cannan, "Enterprise Culture," 10 and 7.

17. *Times Higher* 27/2/1998.

18. Dominelli and Hoogvelt, "Globalisation, Technocratization," 48.

19. Dominelli and Hoogvelt, "Globalisation, Technocratization," 50.

20. Dominelli and Hoogvelt, "Globalisation, Technocratization," 52.

21. Kim Thomas, *Gender, Subject*.

22. Romy Borooah, Kathleen Cloud, Subodra Seshadri, T. S. Saraswathi, Jean T. Peterson and Amita Verma, eds., *Capturing Complexity: An Interdisciplinary Look at Women, Households and Development* (London: Sage, 1994).

23. *Times Higher* 12/12/1997.

24. Spain, *Gendered Spaces*, 168.

25. Spain, *Gendered Spaces*, 233.

26. Meg Maguire, "In the Prime of Their Lives? Older Women in Higher Education," in *Breaking Boundaries: Women in Higher Education*, edited by Louise Morley and Val Walsh (London: Taylor and Francis, 1996), 34; Christine Zmroczek and Pat Mahoney, "Women's Studies and Working-Class Women," in *Desperately Seeking Sisterhood: Still Challenging and Building*, edited by Magdalene Ang-Lygate, Chris Corrin and Henry Millsom (London: Taylor and Francis, 1997).

27. Cyril L. R. James, "Black Studies and the Contemporary Student," in *At the Rendezvous of Victory*, edited by Cyril L. R. James (London: Allison and Busby, 1969/1984), 194.

28. Kelly Coate Bignell, "Building Feminist Praxis Out of Feminist Pedagogy: The Importance of Students' Perspectives," *Women's Studies International Forum* 19, no. 3 (1996): 315–25; Sue Jackson, "Safe Spaces: Women's Choices and Constraints in the Gendered University" (paper presented to Women's Studies Network Conference: Gendered Space, July 1998).

29. Janice Moulton, "A Paradigm of Philosophy: The Adversary Method," in *Women, Knowledge and Reality: Explorations in Feminist Philosophy*, edited by Ann Garry and Marilyn Pearsall (London: Routledge, 1996).

30. Jackson, "Safe Spaces."

31. Ruth Holliday, Gayle Letherby, Lesli Mann, Karen Ramsey and Gillian Reynolds, "A Room of Our Own," in *Making Connections: Women's Studies. Women's Movements. Women's Lives*, edited by Mary Kennedy, Cathy Lubelska and Val Walsh (London: Taylor and Francis, 1993).

32. Gayle Letherby and Jen Marchbank, "To Boldly Go: Safe Spaces and Gendered Places in Feminist Research" (paper presented to Women's Studies Network Conference: Gendered Space, July 1998).

33. White, "Student Needs," 8.

34. Skeggs, *Class, Gender*.

35. White, "Student Needs," 6.

36. Heidi S. Mirza, "Black Women in Education: A Collective Movement for Social Change," in *Black British Feminism: A Reader*, edited by Heidi Saffia Mirza (London: Routledge, 1997), 276.

37. Pascall and Cox, "Education, Domesticity," 31.

38. Kalwant Bhopal, *Gender, Race and Patriarchy: A Study of South Asian Women* (Aldershot: Ashgate, 1997).

39. Uma Narayan, "Contesting Cultures: 'Westernisation,' Respect for Cultures and Third World Feminists," in *The Second Wave: A Reader in Feminist Theory*, edited by Linda Nicholson (London: Routledge, 1970/1997), 399.

40. bell hooks, *Talking Back: Thinking Feminist—Thinking Black* (London: Sheba, 1989), 98.

41. Karen Davies, *Women and Time: The Weaving of the Strands of Everyday Life* (Aldershot: Avebury, 1990).

42. White, "Student Needs," 10, drawing on McIntosh (1990).

43. Rosalind Edwards, *Mature Women Students: Separating or Connecting Family and Education* (London: Taylor and Francis, 1993).

44. Rosalind Edwards, "Access and Assets: the Experience of Mature Mother Students in Higher Education," in *Women and Social Policy: A Reader*, edited by Clare Ungerson and Mary Kember (London: Macmillan, 1997), 272; Jackson, "Safe Spaces," 3; Mary F. Belenkey, Blythe McVicker, Nancy R. Goldburger and J. M. Tarule, *Women's Ways of Knowing: The Development of Self, Voice and Mind* (New York: Basic Books, 1986), 15.

45. hooks, *Talking Back*, 99.

46. Kate McKenna, "Subjects of Discourse: Learning the Language that Counts," in *Unsettling Relations: The University as a Site of Feminist Struggles*, edited by Himani Bannerji, Linda Carty, Kari Dehli Susan Heald and Kate McKenna (Toronto: Women's Press, 1991), 114, citing Smith (1990).

47. Edwards, "Access, Assets," 274.

48. Jacquelyn W. White and John A. Humphrey, "A Longitudinal Approach to the Study of Sexual Assault. Theoretical and Methodological Considerations," in *Researching Sexual Violence against Women: Methodological and Personal Perspectives*, edited by Martin D. Schwartz (Thousand Oaks, CA: Sage, 1997), 23, citing Koss, Gidycz and Wisniewski (1987), who surveyed 6,200 college and university students on thirty-two campuses in the U.S. and found that 53.8 percent of undergraduate women had experienced sexual victimisation at some time.

49. Edwards, *Women Students*; Pascall and Cox, "Education, Domesticity."

50. Edwards, *Women Students*, 80.

51. Madeleine Gromet, "Conception, Contradiction and Curriculum," in *The Education Feminism Reader, Part III: Knowledge, Curriculum and Institutional Arrangements*, edited by Lynda Stone (London: Routledge, 1994), 150.

52. Louise Morley and Val Walsh, eds., *Breaking Boundaries: Women in Higher Education* (London: Taylor and Francis, 1996).

53. Jackson, "Safe Spaces."

54. Kari Dehli, "Leaving the Comfort of Home: Working through Feminisms," in *Unsettling Relations: The University as a Site of Feminist Struggles,* edited by Himani Bannerji, Linda Carty, Kari Dehli, Susan Heald and Kate McKenna (Toronto: Women's Press, 1991), 62–63.

6

Research through the Prism of Space and Time

"But somehow, when I'm doing it [studying] on my own sometime I feel a bit isolated there [the dining room]. Sometime I actually go into the lounge, sit down, have the television on and just feel more at peace . . . Doing it there I actually get more done, than actually doing it in the dining room on my own. Sometime I go to the study room where I do a bit of typing. What I do, mainly typing there and other times as well I might feel more comfortable doing a piece of work in the bedroom. It's strange. I sort of move about the house depending on how I'm feeling on that particular day."

Respondent Zandra, chapter 10

"Today I didn't want to use the study because I like to move around the house."

Researcher Log, chapter 11

Introduction

THE AIM OF THIS CHAPTER is to discuss the research methodology underpinning the research. Liz Stanley and Sue Wise,[1] citing Sandra Harding, argue that it is important to distinguish between the epistemology (theory of knowledge), methodology (perspective) and method (techniques), involved in research. The chapter is in three parts concerned with each aspect.

Research is not an *"orderly hygienic process."*[2] The researcher's social position, knowledge, beliefs and resources inform the work as it progresses. In particular, space and time are significant critical factors shaping the research approach on an everyday basis. In this chapter I include discussion of the ways in which space and time shaped the research. In chapter 11 I reflect on the ways in which space and time informed the research action through an analysis of the researcher's personal log.

Space, Time and Research Epistemology

In chapter 2 I stressed the need for feminist research approaches which accept the possibility of *"real world research"* to give attention to the process of knowledge production, and, whilst recognising differing and complex axes of inequality, identify central regimes of power shaping women's lives. In this chapter, I argue that a research epistemology that includes the concepts of space and time as dynamic rather than merely contextual, and as qualitative as well as quantitative, facilitates understanding of women's oppression. It does this by maintaining the visibility of the complex macro and micro dynamics of women's experiences. I demonstrate this through a discussion of four concepts underpinning my research:

- women's inequality and agency;
- social divisions, inequality and agency;
- the relevance of both materialist and post-structuralist theory;
- the relativity of knowledge.

Women's Inequality and Agency

In relation to exploring women's inequality, Liz Kelly, Sheila Burton and Linda Regan[3] stress the need for feminist researchers to focus on *"the way women's oppression is structured and reproduced."* This entails the use of a diversity of methods and an emphasis on challenge rather than exploitation.[4] Mary Fonow and Judith Cook[5] argue that feminists should be active to *"guard against misuse of findings"* and to focus on the policy implications of their work. There is a need to uncover the experience of different groups of women, together with the processes giving rise to this experience.[6] Feminists stress the need to focus on transforming the regimes of power that shape women's lives.

How are the concepts of *women's inequality and agency* approached? The visibility of space and time throws light on the way women's inequality is produced and contested. Time as a resource is manipulated through the exercise of power: *"Time disposal is partly determined by the individual, partly by social or legal coercion and partly through their negotiations."*[7] The same is true of space, in that the production of space is an exclusionary/inclusionary process.[8] Through exploring women's space/time negotiations, and through use of the concept *others' time*, areas of inequality are exposed.

Social Divisions

How does a consideration of spatial/temporal factors increase insight into social divisions? In terms of the *equally* significant social divisions of class, "race," ability, sexuality and age, it becomes less possible to generalise the experience of one particular group to another. This is because *"lack of time-space referents encourage universalism and generalisation."*[9] Reference to space and time increases the visibility of social divisions and difference. The divergent routes and pathways associated with different groups become visible.[10] In order to address issues of social division and difference in my research, I have assumed that dominant *spatial/temporal practices and representations* shape women students' lives differently in relation to their social position. Conversely, their social position is produced through spatial and temporal relations.

The Relevance of Both Materialist and Post-Structuralist Approaches

Following the discussion of feminist and sociological theory in chapters 2 and 3, I have adopted an open position with regard to these perspectives in my own research. From materialist perspectives I adopt ideas and concepts concerned with wider relations of power and the materiality of experience. From post-structuralism I adopt ideas and concepts concerned with the relativity of knowledge, the local context of experience and shifting social identities. Using both a quantitative and qualitative approach to concepts of space and time (see below) has facilitated this approach. For example, I have gathered quantitative data regarding women students' housing conditions, income, household, paid employment prior to and during their studies and leisure activities. I have also included open questions to allow students to critically reflect on their experiences and actions and to give visibility to the overlapping spheres they experience. As a result, both the current context and materiality of experience together with the dominant meanings associated with space and time are identified.

The Relativity of Knowledge

How does consideration of spatial/temporal factors facilitate reflexivity in research? In all research it is important to adopt research methodologies that increase the visibility of *"behind scenes of power and control."*[11] How is such power and control manifest in research? Pierre Bourdieu[12] argues that forms of knowledge and system arrangements shape the conceptualisation of research objects and that there is a critical need to break with pre-constructions and to explore the relations producing social facts including the researcher's own social position. Joan Parr and others demonstrate that theoretical assumptions may act as a barrier to understanding women's ordinary lives.[13]

In relation to applying a reflexive approach in the research process, I would argue that it is important to consider three areas. Firstly, the categories of knowledge shaping the research need critical evaluation. This goes beyond considering the explicit theorisations of the individual researcher. It concerns the way in which the social world is generally conceptualised and understood, and the way this in turn may relate to powerful interests. What could be the outcomes of research? Secondly, the specific social position of the researcher should be considered. For example, their academic position, access to research funding, academic discipline, institutional base, and the labour process involved.[14] Thirdly, how do the researcher's personal emotions, values, knowledge and experience shape the research process?[15] I now move on to discuss the spatial and temporal relations of my research and address the research question, *"How have space and time shaped the research?"* I consider: core categories of knowledge shaping the research; spatial/temporal practices of the research; spatial/temporal representations of the research.

Core Categories of Knowledge Shaping the Research

Concepts identified in the research correspond to normative and bureaucratic categories. Three concepts are *women, students* and *higher education.* My selection of the concept of *women* involves a choice to deal with this group of people separately. It arises from a normative tradition that has either treated women as *separate from* and *other to* men, or has failed to distinguish their experience from men's. By adopting such a concept from the outset, I am open to criticism because of the universality of experience that it may imply. However, maintaining a focus on *women* does not involve ignoring the complex of class, "race" and gender relations that shape different women's experiences. Himani Bannerji[16] stresses that *difference* is not *diversity* but is evidence of power. She argues that black and white women share some of the same social relations but may end up in very different places. Nevertheless, a tension persists in my research through the drive to seek out areas of commonality rather than difference in women's experience. This drive is exerted through the core concept adopted.

Another concept from the research is *higher education*. This concept reflects a set of institutionalised practices that are distinguished from the process of learning in general. Research into *higher education* has tended to focus on these institutional practices. My research, however, shifts the focus to areas of household, paid work and leisure as well as higher education. Nevertheless, a tension persists in the research between focusing on *higher education* practices and the broader range of practices women adopt *in pursuit* of higher education or of learning in general. A third concept from the research is *student*. This concept describes an individual's specific relationship to the education system, but as argued previously, women's roles are multiple. The meaning of *student* for women is far more complex and shifting than the normative definition would imply. I myself am simultaneously a student and a lecturer in higher education.

Joan Acker, Kate Barry and Johanna Esseveld[17] discuss the way in which unquestioning acceptance of a feminist frame of reference and its associated bureaucratic categories impeded their research. I believe I have facilitated reflexivity in my own research through making the spatial and temporal relations that shape it visible. The conceptualisations I draw on come from bodies of knowledge about the world that are spatially and temporally distinguished (for example, through the concept of public/private spheres). Reflexivity is sustained in research, through reference to *"reality outside the representation."*[18] It is necessary to consider the wider relations (social, spatial and temporal) from which the research arises. The process of abstracting objects and constructing stationary concepts in research arises where spatial/temporal factors are ignored. Andrew Sayer argues, *"First we tear things out of their context, then forget that context and treat the objects as spaceless, timeless data, and then proceed to wonder how we might explain them, which involves trying to reconstruct some kind of appropriate causal context in the absence of information on their spatio-temporal form."*[19] Consequently, maintaining the visibility of space and time allows us to give visibility to process rather than to static structures. Attention is thereby drawn in detail to both the context and nature of action. I now move on to discuss how space and time directly shaped the research.

Spatial/Temporal Practices of the Research

In chapter 5 I identified key issues in higher education practice. These included the rigidity of temporal rationales in higher education, social divisions and the way in which the academic culture is embedded with concepts of individuality, enterprise and production. I now discuss these issues in relation to the construction of my own research. As I was studying for a doctorate, the dominant temporal rationales of higher education shaped my progress in particular ways. There was not a programme of written assignments to submit to deadline but nevertheless, a fairly rigid time scale shaped the research. It was expected that I would complete certain aspects of the work in a certain order and by a certain stage. There was flexibility as to when each stage should be completed but the order of completion was more problematic, in particular the transfer from MPhil to PhD. This was based on a model of progress in research that assumed students to be working at one level (MPhil) and then upgrading to another level (PhD). The model of research informing the transfer was of empirically based scientific research of a deductive nature; the assumption being that at doctoral level, more thorough testing of hypotheses and more analysis would take place than at MPhil level. My own research did not follow this model as I shifted between the literature, research data and analysis throughout the process. The dominant research practice in the institution, in particular, the centralised procedures for upgrade created uncertainty, being based on scientific research protocol.

Individuality, enterprise and production are themes that are embedded in academic culture (see chapter 5). It is perhaps in relation to research activities that these themes are most stressed. The dominant practices of research came through the benchmarks for work at doctoral level; for example, originality, independence, critical depth and so forth. All these concepts are open to contest. Work is perceived to be original when it is in fact closely linked to the academic discourses it attempts to refute.[20] The parameters of academic discourse therefore are quite narrow. Any attempt to break from the adversarial approach creates uncertainty. This was so in my case, where I drew on a range of somewhat conflicting ideas. How was it justifiable to simultaneously adopt materialist and post-structuralist ideas and an inductive and deductive approach (see below, "Time, Space and Research Methodology")? Although working in isolation, the individual researcher in fact collaborates in many different ways. S/he draws on shared knowledge and ideas from previous research, from colleagues, from supervision and from respondents.

Spatial/Temporal Representations of the Research

The *physical landscape* of the research was the institution where I was a student. But in addition the research was carried out at the institution where I worked and at my home. Working through two institutions was complex, particularly when the research institution also had a split campus and libraries. I ultimately abandoned use of that institution and its resources, other than supervision. I decided it was easier to get books from my workplace and to study at home. At one stage during the research, my mobility difficulties led to home visits from my supervisor. The *social landscape* of the research was partially constructed through the academic disciplines available, which represented different areas of social knowledge. The key issue creating uncertainty was the multidisciplinary approach I had adopted. I had to decide within which academic discipline to situate myself. Should I site myself within education studies? In fact my focus was less on teaching and learning than on women's everyday lives. Perhaps cultural studies, sociology or social policy were more appropriate bases? I ultimately chose to site myself in leisure studies because of the feminist orientation of my supervisor who was based there. At a later stage and as a result of supervision I came to realise the very close connections between women's pursuit of leisure and their pursuit of higher education. Both may be conceptualised as seeking *time for self*. The leisure reading proved useful in highlighting the ways in which women conceptualise *time for self*. However, the decision to situate myself in leisure studies led to uncertainty about my coverage of certain issues in the research. For example, were higher education theory and the sociology of space and time being adequately addressed? The decision to situate myself where I did shaped the direction of the research, by drawing attention to certain aspects of women students' experience, but perhaps closing the door on other aspects.[21] The *emotional landscape* of the research I experienced was related to memories of the past and ongoing struggles over space and time. My choice of research focus and the way the research progressed were clearly mediated by this experience (see more detailed discussion of the area in chapter 11).

The *spatial/temporal practices and representations* of research shaped my progress in particular ways. A focus on the spatial/temporal encourages a focus on the *process* of constructing research and gives visibility to the wider relations of knowledge through which research occurs. An essential part of the research process therefore is the need to critically reflect on that process. This I have done through keeping and analysing a personal log during the research. In chapter 11 I critically reflect on the research process. I analyse my role as researcher, drawing on my written log from October 1998 to June 1999, when the main part of the data gathering and analysis took place.

The Epistemology

The core elements of the epistemology underpinning the research are as follows. I question the categories of knowledge underpinning the research through critical reflection on the wider social, spatial and temporal relations which shaped the research in progress and by listening closely to respondents. I draw on concepts concerned with social, spatial and temporal relations to identify the wider material context of women's experience and differences between women. I draw on women's own accounts of their experience and give attention to the narratives of different women, focusing on their thoughts, feelings and action in space and time. Rather than focusing on structures as static, I focus on social relations and on processes of social change. For example, *higher education* is conceptualised as a set of relationships and a process that extends beyond the boundaries of any particular institution. In terms of concepts of social position and social identity, social position is considered to be space/time specific. Social identity shifts in relation to dominant classifications, personal experiences and researcher's views.[22] For example, in this research I conceptualise differences in relation to "race" or ethnicity in several ways. I do not use either of these terms to classify women because they lack specificity and meaning. However I do use black/white in my descriptions of women, although I am aware of the way broad references to colour homogenise differences within and between groups. I use the term following Paul Gilroy's discussion. As skin colour generates different social responses then the language of colour cannot be completely abandoned. The language of colour has political significance.[23] In addition, I identify black women by geographical heritage, a spatial identification that they, in the main, chose (see chapter 7).

Space, Time and Research Methodology

Quantitative or Qualitative?

Research methodologies have been characterised through various orthodoxies. Debate has focused on the strengths and limitations of quantitative and inductive versus qualitative and deductive approaches. Approaches that draw on quantitative methods have been criticised as *positivist*, that is, that they develop and test subjective hypotheses and draw particular *truths* from these. It is mistaken to automatically characterise quantitative approaches in this way.[24] Quantitative approaches are not inherently reactionary.[25] However, Jayaratne and Stewart go on to argue, much quantitative work has been characterised by *"selection of sexist and elitist research topics . . . biased design . . . exploitative relations with the researched . . . illusions of objectivity . . . overgeneralisation . . . inadequate data dissemination and use."* There is a case to be made for quantitative approaches that provide *"better science not no science"* including awareness of the *"metaphysical commitments of the researcher."*[26]

Inductive or Deductive?

In terms of qualitative research, another orthodoxy has dominated discussion. This is that the purest form of research should be uncluttered by prior theorisations, that is, it should be inductive rather than deductive. Hypotheses should not be imposed in advance, rather theory should emerge in the process of research. This is the central tenet of the *grounded theory* approach,[27] an approach that has been attractive to feminist researchers in particular because they challenge dominant knowledge of women and generally believe that the perspectives of oppressed groups will give a clearer account of social reality than dominant regimes of knowledge which are distorted through the interests of particular social groups. Although there are strengths in the grounded theory approach, a critical qualification is that *"inevitably theory precedes research."*[28] Even where every effort is made to give centrality to the voice of the research respondent in gathering and analysing data, cultural constructions shape the process.[29] Liz Stanley and Sue Wise[30] argue that the idea that research can be wholly inductive or deductive is simplistic. Research does not follow a linear model in this way.

I have adopted a research methodology in the belief that no research is solely deductive or inductive and that both qualitative and quantitative approaches have strengths. For example, I use an approach that makes explicit the prior theorisations and commitments of the researcher but also remains grounded in the experience of women. In my research I am explicit about prior theorisations but keep these open enough to allow reflexivity. I would argue that the concepts used are vehicles for reflexivity because they draw attention to spatial and temporal factors and create the space for social differences to emerge, thus avoiding generalisation.

Operationalising Space and Time in the Research Methodology

In chapter 3 I argued that traditionally, where space and time have been considered, the quantitative approaches of mapping and measuring have taken precedence. Quantitative approaches to researching space and time are represented in the work of Daphne Spain[31] and Patricia Hewitt.[32] These writers are concerned to expose processes of exclusion and inequality. A tradition within feminist research of exploring the spatial sexual segregation of women is represented in the work of Spain (see chapter 4). She draws on a large amount of anthropological data and "maps" the physical segregation of women in the spheres of home, paid work and education. She links this mapping exercise to a discussion of the social stratification of women. A similar approach is represented in the work of Hewitt (see chapter 4), but in this case the focus is on physical time rather than space. Hewitt evaluates women's shifting patterns of clock time use in relation to paid work and housework and links this discussion to recommendations for fairness at work. Both the above writers do far more than quantify space and time in women's lives but their overall approaches are essentially quantitative, drawing on concepts of space and time which are open to mapping and time use surveys. Both writers also draw on qualitative research to a lesser extent. The strength of their approach is that a wealth of empirical data is uncovered regarding contemporary and historical *spatial and temporal practices* in relation to women. For instance, the physical segregation of women in different communities and women's paid and unpaid labour.

A variety of techniques have been used to gather quantitative data on space and time including diaries, logs, mapping exercises, interviews and observations. The principal problem besetting these is the difficulty of quantifying concepts that have both a quantitative and qualitative character. Saraswathi argues that time diaries and recall records notoriously underestimate time use.[33] In addition, Karen Davies[34] argues that time use studies measure *"male mechanical time"* and are not able to address the *"patchwork quilt"* of women's time use, deflating both simultaneous work and emotional work. In relation to space, snapshots of spatial practice (e.g., sexual segregation) may be revealed through mapping exercises, but the production of space as a process is less visible. There are limitations to quantitative approaches that may conceal:

- Space and time as qualitative, for example the meanings which different groups attach to time and space.
- Time as event based rather than clock based.[35] Individuals may carry out actions simultaneously,[36] for example, the tasks of paid work and childcare. Shelton,[37] whose research approach to women, men and time use was quantitative, had to exclude child care from her analysis of household labour because of the difficulties in quantifying simultaneous activities.
- Space, time and power, for example, the unequal negotiations around space and time related to social division. These negotiations take place in relation to paid and unpaid work and are shaped through the social divisions of class, colour, geographical heritage, age, gender, sexuality and ability. The financial value of time spent in paid work relates to the social value accorded to particular individuals and to particular work.[38] Research has demonstrated that trade-offs between time and money also take place in the home where for example women may contribute time to the household and men money.[39]
- Space and time as active in social process and not merely contextual.[40]

Qualitative approaches to researching space and time in women's lives are represented in the work of Shirley Ardener et al.[41] and Karen Davies.[42] Ardener draws together the work of a number of feminist researchers studying women and space.[43] They draw on observation, participation, archival analysis and interviewing. Their conceptualisation of space is both quantitative and qualitative. Their research approach is essentially qualitative but involves quantitative techniques. Karen Davies also adopts a qualitative methodology to explore women and time. She argues that a multiple theorisation of time (including both quantitative and qualitative concepts) facilitates research, for example, clock time may be measured, but the values associated with time may not.[44] She draws on concepts related to women's multiple experience of time, *others' time* and the values and meanings connected to time. This approach to researching space and time is qualitative and focuses on experiences, feelings and conceptualisations. It is explicit about the prior theoretical assumptions and thematic frameworks being drawn on. Davies draws on a feminist phenomenological approach developed by Smith who argues that we *"should never lose sight of women as actively constructing as well as interpreting, the social processes and social relations which constitute their everyday realities."*[45] Davies uses thematically structured in-depth repeat interviews with women, where the themes are *"determined in advance."* She also uses grounded theory, allowing *"women to evolve as subjects"* in the research. This allows her to remain perceptive to their world-views. She likens her approach to painting a picture of women's lives, which includes their views, personal relationships, past experiences, hopes and descriptions of events.[46]

My primary approach is along the lines of the work developed by Davies and Ardener et al. I also include explicit consideration of the emotional life of women students. The recognition of the significance of emotion in social analysis has led to important methodological developments. Mary Fonow and Judith Cook[47] argue, *"This aspect of epistemology involves not only acknowledgement of the affective dimension of research, but also recognition that emotions serve as a source of insight or a signal of rupture in social reality."*[48] Joan Parr[49] discusses the relevance of considering emotion in her research with mature women students in higher education. She argues that women's choices to return to study are often linked with traumatic life events: *"The links were not always this clear, but education seemed to be a vehicle which the women were using to deal with some of the consequences of their experiences."* Following from this, I felt it important that my research into women, space and time also include consideration of emotion as a form of action and as a part of space/time negotiations.

Having reviewed some approaches to research which draw on concepts of space and time, it can be seen that rather than being opposed, qualitative and quantitative approaches are in fact closely related. Quantitative techniques can be applied in the pursuit of qualitative data and reflexivity may be applied in the pursuit of quantitative data. Following Jayaratne and Stewart, it could be argued that as all social facts are interpreted and constructed in particular ways through particular systems of knowledge, then in essence all research is qualitative. There is also no simple opposition between inductive and deductive approaches. A basic issue is whether research draws on concepts of space and time that are in themselves quantitative or qualitative. As Barbara Adam[50] argues, it is not necessary to abandon quantitative conceptualisations of space and time completely but it is necessary to be explicit about the concepts being drawn on.

The Methodology

My own research methodology therefore draws on the approaches developed by Shirley Ardener and Karen Davies together with approaches that include consideration of emotion. I also use some quantitative techniques to develop a profile of women students. I felt it was appropriate to explore the ways in which women integrate higher education into their lives through their own accounts of their experiences, perceptions and emotions in relation to the process. I draw on a feminist qualitative methodology.[51] As I have argued, there are limitations to research approaches that attempt to quantify space and time, even where they attempt to quantify process rather than outcome. Nevertheless, quantitative techniques enable *snapshots* of current dominant *spatial/temporal practice* to become visible.

The methodology for my research contains the following elements. It is feminist and qualitative, being concerned with the complex relationship between processes related to gender, space, time, structures of power and personal experience. As in Davies, the approach conceptualises women as both constructing and interpreting their social world. The research is explicit about the prior theoretical frame which includes both quantitative and qualitative conceptualisations of space and time, concepts of gender and social division and concepts of higher education. However it is responsive to and respectful of participants' voices, frames of meaning and emotions throughout the stages of data collection, analysis and writing up, drawing on grounded theory techniques to achieve this.[52] The research is consciously reflexive, being concerned with the ways in which the spatial/temporal relations I was subject to including the knowledge base, my social position and my experiences and emotions informed the research process at each stage. The research is positional in that social position and social identity are explored through critical reflection on dominant representations, respondents' perspectives and researcher perspectives.

Space, Time and Research Methods

The Wider Sample

Forty-six women from two degree programmes and in their final year of studies were selected. The students were based at the community college where I worked and were studying for degrees in the social sciences. They reflected a variety of backgrounds and experience in terms of their age, geographical heritage, colour, households, place of birth, accommodation, class background, health, sexuality and dis/ability. They were involved in two different academic pathways. I was their tutor in one module. These women were given the semi-structured questionnaire and students' reflective log.[53] This sample was selected for the following reasons:

- the diversity of their backgrounds facilitated analysis of the breadth as well as the depth of women students' experiences;
- the continuum of women's experiences throughout and soon after the period of their studies was accessible to research (timescale three/four years);
- my active involvement with these students as tutor (over one year) as well as researcher facilitated the research.

The Smaller Sample for Interview[54]

Seventeen women were selected for follow up interviews after they had completed their final year studies. I selected this smaller sample in order to gain insight into the implications of the following axes of women students' experience:

- age;
- living with and without dependent children;
- colour;
- geographical heritage;
- type of residence;
- sexuality;
- income;
- living with partners and living alone;
- caring responsibilities;
- health and impairment issues.

The majority of women interviewed for the pilot study came from similar backgrounds to me in terms of age, mother status, sexuality, geographical heritage and colour. Although a tutor, I was nevertheless on a fairly similar critical plane to these students. I therefore selected the second sample for the doctoral research interview in the light of this. Six of the seventeen interviewed were black women (three of African Caribbean and three of South Asian heritage), six were women under twenty-two years (that is school leavers on entry). Nine women had no children. I interviewed women from a range of backgrounds to ensure that issues of social position were highly profiled in the research and to provide an in-built challenge to any assumptions that arose from my own experiences as a white mature mother student. In the analysis I drew on both interviews and questionnaires to ensure that issues of social position were adequately covered.

Researcher's Personal Log

I kept a log identifying critical events during the period of the research. The aims were threefold. Firstly, to facilitate reflexivity throughout the research process. Secondly, to gather data on the *spatial/temporal practices and representations* shaping the research and on my actions in the spheres of home, paid work, research and leisure and thirdly, to provide a rich source of data regarding my work with students. The log is discussed in chapter 11.

Semi-Structured Questionnaire[55]

This was to develop a profile of participants in terms of their positioning, the demands on their time, their accommodation and the places where they study (quantitative concepts of space/time). It was also a means of gathering data on women's actions (including perceptions and feelings) in relation to studying (qualitative concepts of space/time). It was used to develop *snapshots* of the dominant *spatial and temporal practices* women students are subject to in the spheres of home, paid work, community/leisure and higher education. In addition it was a means of identifying a sample of women of different backgrounds and experiences for further interviews. The questionnaire was given to the whole sample. Open and closed questions were structured in relation to the spheres of household, housing, income, paid work, leisure/community and identity. These questions explored current circumstances and transitions during the period of their academic studies. It was given out in December 1998 and students completed it in class. This was to facilitate return of the questionnaires and to limit interference of the research with students' time for study and other activities. A questionnaire can be an efficient way of accessing data in a short period of time. It may also give students a greater degree of control over what they choose to include/exclude than an interview. Regarding issues of identity, whilst giving leeway to women to self describe, I gave indicators along conventional lines of status.[56] It was interesting to see how the women related differently to these conventions. The primary analysis of this questionnaire is in chapter 7.

Students' Reflective Log[57]

This tool was to explore the individual action involved in studying, including critical events/turning points in relation to studying, and the ways that women actively engage with the spatial/temporal relations of home, paid work, leisure/community and higher education in order to study. Linda Bell[58] discusses the limitations of diaries and logs in exploring time use. She argues that it is difficult to distinguish what people *"say, do and say they do"* in diaries. It is also difficult to differentiate between clock time and process time as the diary may encourage respondents to conceptualise time as *clock time*. For example, *"You can count time spent as a teacher but teacher time is woven through other time."*[59] Diaries are also vehicles for private thoughts and feelings. They are *"versions of reality"* which might leave out emotions when for public consumption, but include them when written for self.[60] Diaries are also limited by recall.

The reflective log was designed to encourage students to give accounts of studying which included both quantitative and qualitative concepts of space and time. This included the space and time they used to write an essay, critical events that took place in the process of achieving this and their actions (thoughts, feelings and behaviour). It was initially used in the pilot research where it proved effective. It was given to the wider sample of women students. I asked them to reflect on the process of producing a particular essay immediately after they had handed it in for marking. They were asked to write a log relating to this essay. The log concerned the period from beginning to plan the essay to handing the work in. Questions were structured in relation to the spheres of housework, paid work, leisure and relationships. Women gave accounts of experiences, perceptions and feelings in relation to the creation of study space and time for the piece of work. I asked students to write the log in May 1999 in a module evaluation session. The contents of the logs are mainly analysed in chapter 10.

Semi-Structured Interviews[61]

This was to explore in detail the experiences, perceptions and emotions (actions) of women students in relation to the process of integrating higher education into their lives. This involved deepening understanding of students' *"complex temporalities of contextual being"*[62] in relation to higher education and their everyday lives. It was a means of gathering case study material. The interviews facilitated the development of *landscapes* of women's experience at a physical, social and emotional level; and revealed the ways that women actively engaged with the spatial/temporal relations in different spheres of everyday life in order to study. I interviewed the seventeen women from the smaller sample. They were asked to reflect on the process of studying in higher education shortly after they had completed their final year. The themes of space and time informed the interview design but did not constrain students from articulating their experiences in ways they chose. In depth, thematically structured interviews encouraged women to articulate their experiences and actions (including conceptualisations and emotions) in relation to studying and explored the links between higher education and significant life events.[63] These interviews were conducted where women felt most at ease and taped when they consented.

I introduced the themes of space and time in the interviews. This was despite the fact that Karen Davies[64] argues that it is important not to ask directly about time in qualitative research. In relation to her own research she felt this would constrain women to give accounts of their experience that drew on dominant perceptions of time (clock time). Barbara Adam[65] argues that people's everyday lives are influenced by the *"dominant timescape"* in which *"actors feel pride"* therefore direct questioning about space and time may lead to narrow accounts of women's experience. Questions that directly address everyday life may be a more fruitful source of information on the ways women actually experience and conceptualise space and time. However, in my case I felt that women's perceptions and experience in relation to space and time would be a fruitful source of information in relation to their everyday lives. Following Eileen Green and Diane Woodward,[66] who researched women's experiences of leisure, I asked direct questions about space and time. Green and Woodward chose not to ask direct questions about leisure because of the difficulty women have in conceptualising this. They found that questions about time use were more helpful in uncovering women's concepts and experiences of leisure. In the interviews I facilitated descriptions of space, time and studying which were both quantitative and qualitative in nature. My approach was to encourage students to draw on a wide range of conceptualisations of space and time (including physical, social and emotional).

Transcribing

I transcribed interviews verbatim and made notes regarding various interjections. Liz Kelly[67] developed a code to transcribe emotional expression as well as what was said, in relation to both researcher and respondent. I did not develop a code for the transcripts but noted down what I considered to be significant detail. I wrote brief case studies in my log that addressed issues of setting, my relationship with the student concerned, the progress of the interview and feelings I had at the time of the interview. Transcription is a very time consuming task and it was necessary to weigh up what was desirable with what was realistic at the time.

Analysis

Several researchers have stressed the importance of maintaining reflexivity in the research analysis.[68] Ribbens and Edwards discuss the dilemmas involved in *letting go* of academic knowledge in order to be true to the respondent's voice. They focus on the power of the *narrator* in shaping the research. It is also important to examine the ways the researcher's frame of reference relates to and interacts with the respondents' frame of reference. Following Acker, Barry and Essevold, "*shifting concepts*" should be focus of research. This is the way to *"understand how larger structures penetrate the lives of individuals and the way individuals reproduce and undermine structures."*[69]

In the case of my research, I was explicit about my prior theorisations, and read, codified and analysed data drawing on concepts previously discussed. I attempted to answer questions concerning women's actions (including thoughts and emotions), in relation to integrating higher education into their lives. I explored dominant *spatial/temporal practices and representations* in the spheres of paid work, home, community/leisure related to this process; and the effect of/production of social position. I examined my own position and action as researcher. I read questionnaires and analysed them firstly in relation to the questions I had asked. I then read and re-read the interviews and began to develop themes arising from women's own accounts. I then returned to the questionnaires to develop these themes. Ultimately I came up with the secondary space/time concepts (see chapter 3) including *snapshot, landscape, rhythm and movement*. These arose from a detailed reading of women's accounts of experience and were grounded in that experience. I drew on these concepts to analyse the qualitative data and my research log.[70]

Ethical Issues

As I was academically involved with the students who participated in the research, there were a number of ethical implications. Would students feel obliged to co-operate? In what ways would the accounts of experience that they shared with me be influenced by my role? In what ways would the way I formulated the research be shaped by my role? Would the research interfere with their degree studies? Nevertheless I feel there were particular strengths for the research in building on tutor/student links. My contact with the students enriched the research as my role was specifically focused on their learning and the difficulties they faced in association with this. See chapter 11 for fuller discussion.

I responded to the ethical issues by issuing clear research statements with questionnaires and at the outset of interviews that outlined aims, discussed issues of consent and confidentiality and were explicit about the ways in which data would be used. For example, I explained that it would have no bearing on their assessment.[71] I assured women that I would not look at any completed questionnaires or the reflective log until they had finished their final exams. I conducted in-depth interviews after they had completed their final year assessment. I assured anonymity. Pseudonyms are used throughout and some identifying detail changed, for example, places of work. I created space within the timetable to facilitate their work on the reflective log and semi-structured questionnaire. This was negotiated with other tutors and agreed on. Students were under pressure to complete assessment tasks at home so it was felt this would minimise the interference of the research with their academic work and guarantee their return. I addressed the concerns of non-participant students who may have felt disadvantaged. I asked open questions to enable students to take the lead in interviews as far as possible. I kept a reflective log myself, where, for example, potential conflicts of interest were discussed.

Conclusion

Above I have outlined key elements of my research approach, considering critical issues in relation to the epistemology, methodology and methods I adopted. In each case I have discussed the relevance of the concepts of space and time and the ways in which these are applied. The use of such concepts has, I hope, facilitated the research in three ways. Firstly, the research draws on a wide knowledge base, including materialist and post-structuralist theory, and a wide range of quantitative and qualitative conceptualisations of space and time. Secondly, it contains detailed accounts of women students' actions in their social, spatial and temporal context. Thirdly, it contains a reflexive analysis that is intended to identify the strengths and limitations of the findings by revealing the ways in which the research was constructed. In the following chapters I move on to discuss and analyse the research findings.

Notes

1. Liz Stanley and Sue Wise, "Method, Methodology and Epistemology in Feminist Research Processes," in *Feminist Praxis*, edited by Liz Stanley (London: Routledge, 1990), 26.

2. Liz Stanley and Sue Wise, "Feminist Research, Feminist Consciousness and Experiences of Sexism," in *Beyond Methodology: Feminist Scholarship as Lived Research*, edited by Mary M. Fonow and Judith A. Cook (Bloomington: Indiana University Press, 1991), 266.

3. Liz Kelly, Sheila Burton and Linda Regan, "Researching Women's Lives or Studying Women's Oppression? Reflections on What Constitutes Feminist Research," in *Researching Women's Lives from a Feminist Perspective*, edited by Mary Maynard and June Purvis (London: Taylor and Francis, 1994), 29.

4. Kelly, Burton and Regan, "Women's Lives," 28.

5. Mary M. Fonow and Judith A. Cook, "Back to the Future," in *Beyond Methodology. Feminist Scholarship as Lived Research*, edited by Mary M. Fonow and Judith A. Cook (Bloomington: Indiana University Press, 1991), 5.

6. Sherry Gorelick, "Contradictions of Feminist Methodology," in *Race, Class and Gender: Common Bonds, Different Voices*, edited by Esther Ngan-Ling Chow, Doris Wilkinson and Maxine B. Zinn (London: Sage, 1996).

7. Karen Davies, *Women and Time: The Weaving of the Strands of Everyday Life* (Aldershot: Avebury, 1990), 37–38, citing Haines (1987).

8. Henri Lefebvre, *The Production of Space* (Oxford: Blackwell, 1991).

9. Shulamit Reinharz with Lynn Davidman, *Feminist Methods in Social Research* (New York: Oxford University Press, 1992), 117, citing Daniels.

10. Nira Yuval Davis, *Gender and Nation* (London: Sage, 1997), 86; Beverley Skeggs, *Formations of Class and Gender: Becoming Respectable* (London: Sage, 1997), 94–95.

11. Derek Layder, *New Strategies in Social Research* (Cambridge: Polity, 1993), 249.

12. Pierre Bourdieu, Jean-Claude Chambordon and Jean-Claude Passeron, *The Craft of Sociology. Epistemological Preliminaries* (New York: Walter de Gruyter, 1968/1991), 249.

13. Joan Parr, "Theoretical Voices and Women's Own Voices: The Stories of Mature Women Students," in *Feminist Dilemmas in Qualitative Research: Public Knowledge and Private Lives*, edited by Jane Ribbens and Rosalind Edwards (London: Sage, 1998), 92; Melanie Mauthner, "Bringing Silent Voices into a Public Discourse: Researching Accounts of Sister Relationships," in *Feminist Dilemmas in Qualitative Research: Public Knowledge and Private Lives*, edited by Jane Ribbens and Rosalind Edwards (London: Sage, 1998); Tina Miller, "Shifting Layers of Professional, Lay and Personal Narrative: Longitudinal Childbirth Research," in *Feminist Dilemmas in Qualitative Research: Public Knowledge and Private Lives*, edited by Jane Ribbens and Rosalind Edwards (London: Sage, 1998); Ann Tait, "The Mastectomy Experience," in *Feminist Praxis*, edited by Liz Stanley (London: Routledge, 1990); Jane Ribbens and Rosalind Edwards, "Living on the Edges: Public Knowledge, Private Lives and Personal Experience," in *Feminist Dilemmas in Qualitative Research: Public Knowledge and Private Lives*, edited by Jane Ribbens and Rosalind Edwards (London: Sage, 1998), 2.

14. Liz Stanley, "Feminist Praxis and the Academic Mode of Production," in *Feminist Praxis*, edited by Liz Stanley (London: Routledge, 1990).

15. Reinharz with Davidman, *Feminist Methods*, 67.

16. Himani Bannerji, "But Who Speaks for Us? Experience and Agency in Conventional Feminist Paradigms," in *Unsettling Relations: The University as a Site of*

Feminist Struggles, edited by Himani Bannerji, Linda Carty, Kari Dehli, Susan Heald and Kate McKenna (Canada: Women's Press, 1991), 86.

17. Joan Acker, Kate Barry and Johanna Esseveld, "Objectivity and Truth. Problems in Doing Feminist Research," in *Beyond Methodology. Feminist Scholarship as Lived Research*, eds. Mary M. Fonow and Judith A. Cook (Bloomington: Indiana University Press, 1991), 144.

18. Barbara Adam, *Timescapes of Modernity: The Environment and Invisible Hazards* (London: Routledge, 1998), 6.

19. Andrew Sayer, *Method in Social Science: A Realist Approach* (London: Routledge, 1992), 83–84, 146–47.

20. Janice Moulton, "A Paradigm of Philosophy: The Adversary Method," in *Women, Knowledge and Reality: Explorations in Feminist Philosophy*, edited by Ann Garry and Marilyn Pearsall (London: Routledge, 1996).

21. The *social landscape* of research is also shaped through financial negotiations. Funding is a critical issue shaping research outcomes. Funding bodies often require accountability and sometimes the independence of research may be compromised. In my case, I was in the main self-financing and therefore my main accountability was to the students involved in the research. I had a little financial help from my employers, though they put no constraints on the research.

22. Linda Alcoff, "Cultural Feminism versus Post Structuralism: The Identity Crisis in Feminist Theory," in *The Second Wave: A Reader in Feminist Theory*, edited by Linda Nicholson (London: Routledge, 1988/1997).

23. Paul Gilroy, *Small Acts: Thoughts on the Politics of Black Culture* (London: Serpent's Tail, 1993), 14.

24. David Silverman, *Interpreting Qualitative Data* (London: Sage, 1993).

25. Toby Epstein Jayaratne and Abigail V. Stewart, "Quantitative and Qualitative Methods in the Social Sciences: Current Feminist Issues and Practical Strategies," in *Beyond Methodology: Feminist Scholarship as Lived Research*, edited by Mary M. Fonow and Judith A. Cook (Bloomington: Indiana University Press, 1991), 86.

26. Katheryn Pyne Addelson, "The Man of Professional Wisdom," in *Beyond Methodology: Feminist Scholarship as Lived Research*, edited by Mary M. Fonow and Judith A. Cook (Bloomington: Indiana University Press, 1991), 30.

27. Developed by Glaser and Strauss and discussed by Stanley and Wise, "Feminist Research," 266.

28. Stanley and Wise, "Feminist Research," 267.

29. Pam Alldred, "Ethnography and Discourse Analysis. Dilemmas in Representing the Voices of Children," in *Feminist Dilemmas in Qualitative Research: Public Knowledge and Private Lives*, edited by Jane Ribbens and Rosalind Edwards (London: Sage, 1998), 155.

30. Stanley and Wise, "Feminist Research," 265.

31. Daphne Spain, *Gendered Spaces* (Chapel Hill: University of North Carolina Press, 1992).

32. Patricia Hewitt, *About Time: The Revolution in Work and Family Life* (I. P. P. R. London: Rivers Oram Press, 1993).

33. T. S. Saraswathi, "Women in Poverty Contexts," in *Capturing Complexity: An Interdisciplinary Look at Women, Households and Development*, edited by Romy Borooah, Kathleen Cloud, Subodra Seshadri, T. S. Saraswathi, Jean T. Peterson and Amita Verma (London: Sage, 1994).

34. Karen Davies, *Women and Time: The Weaving of the Strands of Everyday Life* (Aldershot: Avebury, 1990), 45.

35. John Hassard, ed., *The Sociology of Time* (London: Macmillan, 1990).

36. Sharon Nickols and Kamala Srinivasan, "Women and Household Production," in *Capturing Complexity: An Interdisciplinary Look at Women, Households and Development*, edited by Romy Borooah, Kathleen Cloud, Subodra Seshadri, T. S. Saraswathi, Jean T. Peterson and Amita Verma (London: Sage, 1994).

37. Beth Anne Shelton, *Women, Men and Time: Gender Differences in Paid Work, Housework and Leisure* (New York: Greenwood Press, 1992).

38. Wendy Bottero, "Clinging to the Wreckage? Gender and the Legacy of Class," *Sociology* 32, no. 3 (August 1998): 469–90.

39. Shelton, *Women, Men*, 2.

40. John Urry, "Sociology of Time and Space," in *The Blackwell Companion to Social Theory*, edited by Bryan S. Turner (Oxford: Blackwell, 1996).

41. Shirley Ardener, ed., *Women and Space: Ground Rules and Social Maps* (Oxford: Berg, 1993).

42. Davies, *Women, Time*.

43. Ardener, *Women, Space*, 46. Their methods are ethnographic, aiming: to document the lives and activities of women; to understand the experience of women from their own point of view; to conceptualise women's behaviour as an expression of social context (Reinharz with Davidman, *Feminist Methods*, 51).

44. Davies, *Women, Time*, 9. Her stated aim is to *"develop a theoretical understanding of time from a feminist perspective and to analyse a specific group of women's everyday lives using this particular perspective."*

45. Stanley and Wise, "Method, Methodology," 34.

46. Davies, *Women, Time*, 62 and 63.

47. Fonow and Cook, "Back, Future," 3.

48. Fonow and Cook, "Back, Future," 9, citing Cook 1988.

49. Parr, "Theoretical Voices," 99.

50. Barbara Adam, *Timewatch: The Social Analysis of Time* (Cambridge: Polity Press, 1995).

51. Stanley, "Feminist, Praxis"; Mary Maynard and June Purvis, *Researching Women's Lives from a Feminist Perspective* (London: Taylor and Francis, 1994); Jane Ribbens and Rosalind Edwards, "Living on the Edges. Public Knowledge, Private Lives and Personal Experience," in *Feminist Dilemmas in Qualitative Research: Public Knowledge and Private Lives*, edited by Jane Ribbens and Rosalind Edwards (London: Sage, 1998).

52. Amanda Coffey and Paul Atkinson, *Making Sense of Qualitative Data: Complementary Research Strategies* (London: Sage, 1996); Parr, "Theoretical Voices."

53. See below and appendices 1a and 1b.

54. Appendix 4.
55. Appendix 1a.
56. Appendix 1a, "Semi-structured questionnaire," question 7.1.
57. Appendix 1b.
58. Linda Bell, "Public and Private Meanings in Diaries: Researching Family and Childcare," in *Feminist Dilemmas in Qualitative Research: Public Knowledge and Private Lives*, edited by Jane Ribbens and Rosalind Edwards (London: Sage, 1998).
59. Bell, "Public, Private," 81–82.
60. Bell, "Public, Private," 77.
61. Appendix 1c.
62. Adam, *Timescapes, Modernity*, 11.
63. Parr, "Theoretical Voices."
64. Davies, *Women, Time*, 63.
65. Adam, *Timescapes, Modernity*, 5.
66. Eileen Green and Diane Woodward, "Women's Leisure Today," in *Sociology of Leisure: A Reader*, edited by Charles Critcher, Pete Bramham and Alan Tomlinson (London: E. and F. N. Spon, 1995).
67. Reinharz with Davidman, *Feminist Methods*, 40, citing Kelly.
68. Charles C. Ragin, *Constructing Social Research: The Unity and Diversity of Method* (Thousand Oaks, CA: Pine Oaks Press, 1994), 87; Ribbens and Edwards, "Living, Edges," 16; Parr, "Theoretical Voices."
69. Acker, Barry and Esseveld, "Objectivity, Truth," 149–50.
70. Verification of the research was through contrasting and comparing data from the respective research tools outlined in this chapter
71. Appendix 1.

7

Spatial and Temporal Practices: Frameworks for Action

"... spatial practice under neo capitalism ... daily reality (daily routine) and urban reality (the routes and networks which link up the places set aside for work, 'private' life and leisure)."[1]

Introduction

IN CHAPTERS 7 TO 10 I present and analyse the empirical data drawing on concepts developed in chapter 3 and research questions developed in chapters 4 and 5. This chapter presents data from the semi-structured questionnaire. The analysis focuses on the *spatial/temporal practices* shaping women's experiences in the spheres of paid work, housing and household and leisure/community. In chapter 8 the focus is on data from the interview schedule in relation to the concepts of *spatial/temporal representations*. Chapter 9 explores data in relation to the college and chapter 10, women students as *centres of action*, creating space and time to study. This chapter is in four parts, examining in order:

- spatial/temporal practices, social position and social identity;
- spatial/temporal practices and paid work;
- spatial/temporal practices, housing and household;
- spatial/temporal practices and leisure/community.

In each part I give general background detail for the whole sample, identify areas of similarity and difference between women and discuss critical issues, drawing on individual women's written comments. In this way I construct *snapshots* of women students' experience (see chapter 3, "Spatial/Temporal Concepts Underpinning the Research"). The concept of *snapshot* (in contrast to the concept of *landscape,* which is utilised in chapter 8) is characterised as follows. It provides a static image constructed from a range of quantitative and qualitative data. It offers partial insight into the web of relationships and practices women were involved in at specific times and places in their academic careers. It is based on data drawn from the memories of the respondents and mediated through the researcher's experiences (see chapter 11). Hence, the *snapshots* are constructed selectively. The focus is less on the meanings associated with particular places and activities than on the concrete *spatial/temporal practices* that women experienced throughout their studies, that is, the places, routes, networks and temporal rationales through which women experienced everyday life. Women's daily routines, including the places they lived, worked and socialised[2] were interrupted by their academic studies. Here, I begin to address critical issues arising from this process. In the higher education literature certain areas of action tend to be disassociated from the educational sphere. A woman's home life may be considered as something that is brought to the curriculum as a resource or as a learning barrier. Paid work may be considered as a starting point or an outcome in relation to higher education. Both in fact are centrally meaningful parts of the higher education process (see chapter 5). The research question being addressed is, *"What are the dominant spatial/temporal practices shaping women's everyday experiences whilst studying?"*

It was necessary to quantify some aspects of the research data in order to identify critical material factors shaping women students' experiences and patterns of similarity and difference. However, many aspects of the data were not quantifiable. Consequently, the analysis in this chapter involves both quantitative and qualitative techniques. For example, some issues were explored through closed questions and these findings are presented in tabular form where appropriate, in appendix 2. Other issues were addressed through open questions and I have organised responses thematically.

The forty-six women students involved in the research were in the final year cohort of two degree programmes in the areas of social, community, leisure and health studies.[3] The implications of selecting such women are that the degree routes they are on cater particularly for those seeking careers in health, social care and the leisure industry. Social and health care occupations are commonly identified as women's work.[4] Thus, the sample of women involved in this research was, in general, following typical rather than atypical career routes in terms of gender.

Spatial/Temporal Practices, Social Position and Social Identity

Nira Yuval Davis[5] argues that identities are constructed socially from the interconnections between personal experience and cultural conditions. Space and time are also critical in the construction of social position and identity. Women students' social position and identity were explored in the research in a combination of ways reflecting the processes whereby these are socially produced. For example, dominant practices of social classification inform the research in several ways.[6] However, I was also concerned to discover the ways in which students identified themselves. I asked them to write a short pen picture about what social identities they saw as relevant to themselves. The outcome was a very complex picture of the relations out of which women students develop concepts of themselves and the processes whereby identity is constructed.[7] At this stage I discuss issues related to age, class, ethnicity, income, disability and sexuality. Other identity positions (for example mother status) are discussed in later sections of the chapter.

Age[8]

The forty-six women students who responded to the questionnaire came from a wide range of backgrounds. Just over half of those responding to a question on age were under twenty-five when they began their studies. Nearly half (48 percent) were older mature students and six were in their forties. Three-quarters of the older women lived with children. This was a relatively old cohort of students in terms of traditional entry to higher education. Age and stage in the life cycle were clearly critical factors shaping the actions and experiences of women students. They had a bearing on where women lived, whom they lived with and the way their time was occupied. Clearly the likelihood that women lived with their own children increased with age. This applied to over three-quarters of women in their thirties and forties and less than a quarter of women in their twenties. The likelihood of being a lone parent also increased with age. A quarter of women in their thirties and half of the women in their forties were lone parents.

Age was also a critical factor in shaping women's route into higher education and their motivation to study.[9] Fifty percent said they had come from school and 50 percent from a college of further education. Only a third had traditional academic qualifications[10] on entry, others having studied vocational qualifications or come through access courses. Older women and lone parents were far more likely to have come through non-traditional routes.

Age and stage in the life cycle also influenced women's expectations of higher education. All forty-six women responded to a question asking for their reasons for entering higher education.[11] They gave a number of reasons for their choice. These responses were also age related and included perceptions of future advancement through higher education, in relation to career (57 percent), personal self-esteem and autonomy (40 percent) and/or finance (22 percent). Smaller numbers mentioned wider opportunities in general, family advancement, leisure and social activities. There was a shift in emphasis related to age. For example, five of the six women aged forty to forty-five gave personal biography and life events (such as divorce) as reasons for entering higher education whereas only a quarter of those aged twenty to twenty-five did so. However, in many cases responses were holistic. Women implicitly referred to both life history and their long-term aims. One woman simply said she had come into higher education, *"To make life better for myself and my children"* (Kelly); another, *"To disprove the teacher that said 'Mary will never achieve more than stacking shelves.'"* Higher education was conceptualised by women students as a route of change in material conditions, social position and concepts of self.

Class

The women chose whether to identify themselves in terms of class. I decided against using conventional categorisations of class. Paid work, as an indicator was not viable because of women's student status. Their paid work in most cases was temporary and transitional. The paid work they were involved in as students demonstrated that virtually all of them would fall into the lowest socio-economic position, as they worked in routine or semi-routine occupations.

Sixty-four percent of the women chose to identify in terms of class. Of these, 60 percent said they were working class and 40 percent middle class. Of those identifying as middle class several factors emerged on closer examination. Some identified themselves as middle class because of their male partners' occupation or their parents' occupation. Others identified as middle class because of their own previous paid work and expectations of future paid work. The actual differences between women identifying as middle class and women identifying as working class were quite small, but there were some significant ones. There was no significant difference in type of employment or rates of pay whilst students. However, women who identified as middle class were more likely to have had previous paid clerical or administrative posts. They were also more likely to be owner-occupiers. Those identifying as working class were more likely to say that they had less time for leisure than middle class women. It was interesting to see who did not self identify by class. Only three of the seven black women students identified by class (see below) whereas over three-quarters of white students did so; also none of the lone parents identified by class. Was there less to be gained for some women from such identification?[12] Or were the other identities they related to more visible in public discourse?[13] Most women students viewed higher education as a means to advance themselves, wanting to obtain higher professional/managerial posts in the future, thus a means, although unstated, of moving up the dominant classification system.

Colour and Geographical Heritage

As discussed earlier (chapter 4), I do not identify women students in terms of "race" or ethnicity because these terms are open to wide and contested interpretation.[14] I refer to women in terms of their geographical heritage, as black women themselves did in the main. I also refer to colour, although recognising that this homogenises differences between women. Colour nevertheless generates different social responses and therefore such language cannot be completely abandoned.[15] The concepts of black/white have political significance for many women. As with class, over three-quarters of the white women students identified by colour whereas less than half the black women students did so. Black women were more likely to make a spatial identification by naming a country of heritage. The majority did so whereas this was the case for only 10 percent of white women, most of whom came from outside England. There were no significant differences in patterns of paid work, and hourly pay between black and white women.[16] Young South Asian women were less likely to live in halls of residence. Black women of all backgrounds were more likely to identify a community religion as a source of leisure/community. These issues are further discussed below.

Wider Income

It was not possible to discover women students' actual income. This would have involved gathering detailed information about household income, parental contributions (in cash and kind), negotiations with partners and adult children for board and so forth. For many students this would have been an invasion of privacy, particularly because of their own anxieties in relation to grant and benefit authorities and because the college itself may be perceived by students as sharing information with such authorities. However, it was possible to identify the main sources of women students' income and perceived shifts in income since they became students. Forty-five women responded to questions about sources of current income: 80 percent were in paid work; 87 percent received (a little) grant income and 60 percent loans; 31 percent relied on parental contributions; 31 percent social security benefits and 22 percent lived with a male partner in paid work. Six students had no student grant.[17] All of the women who identified as middle class were in paid work whereas less than three-quarters of those who identified as working class were. Only two of the lone mothers said they received child support.

Women students said they were increasingly financially dependent on others in their households and families. For younger students this dependency was mainly on parents, for older women, on partners. Half of the women were involved in intense or very intense negotiations with parents or partners over income.[18] Older single women (with and without children) had less financial dependency on other adults. This did not mean they were necessarily financially worse off, as financial dependency does not mean that money is always forthcoming.[19] All the black students and lone parents felt they had full control of (albeit limited) income. Access to and control of income was not necessarily viewed as positive, *"Access to all the money and control of all money—total financial responsibility—is a huge burden"* (Katrina). Women also distinguished between actual access to and control of money and their feelings about this: *"I don't feel I have equal access to the money as I haven't earned it and in some way my student-grant and loans are family shared and my sort of 'pay-back'!"* (Rian).

Women described general reductions in income as a consequence of studying.[20] Time available for paid work lessened and outgoings associated with studying increased (particularly travel and child-care costs). Women mentioned the impact of the loss of social security benefits (such as Carer's Allowance), problems with the Child Support Agency and problems of budgeting when the student grant was paid infrequently. However, other factors contributed to women students' financial stability whilst studying, in particular the quality of relationships with partners and parents on whom they may have financially depended and changes in household and parental income related to health, redundancy, retirement, divorce and separation: *"Not able to contribute to housekeeping caused tension with lack of money with household. Father off work . . . (health) . . . Stressed, depressed and unable to control his feelings, leading often to outbursts causing rest of household to become upset"* (Barbara).

Illness and Disability

No students referred to themselves as disabled women, possibly because of the continuing negative social connotations associated with this classification. Nearly one-third of the sample said they were affected by illness at some stage during the course. The spaces they lived in seemed to be critical in shaping the health experience of younger women. More than half of those who wrote about ill health lived in halls: *"Mainly usual illness, but due to close contact with others it lasts longer. Sick of hearing coughing etc. all around the corridor"* (Olivia). Over one-quarter were affected by long-term conditions and a third by mental distress: *"Stress of coping with three jobs, college, a child and a house and the separation from a five year relationship made last year and the beginning of this year very difficult. Constantly living in debt is also a source of stress"* (Maggie). One woman identified herself as needing personal assistance. Her personal needs consumed her own time and in addition she received assistance from friends.

Eight women were closely involved with a person with impairment or long term illness and seven gave some personal assistance. Four of these lived in the parental home and the implications of this are followed up in chapter 8. Women gave personal assistance to siblings, parents, grandparents, their own children, neighbours and friends with illness or impairment.

Sexuality

Forty-five percent of the women chose to identify themselves in the questionnaire by sexuality. All of these said they were heterosexual. Only one woman interviewed said she was a lesbian. No student came out as lesbian or bi-sexual in the questionnaire. This raises questions about the limitations of the questionnaire format to raise issues of sexual identity. Issues related to lesbian identity were discussed by several students in interviews (see chapters 8 and 9).

Other Ways of Self Identification

Women responded to the opportunity to self identify with a wide range of descriptions of themselves. Forty-five percent of the women identified themselves in relation to family and kin relationships, mostly as *mothers* and *wives*. Twenty percent identified themselves in relation to other roles and relationships for example as *students* or *friends*. Eighteen percent identified themselves by abilities and innate traits such as *hard working* and *funny*. Ten percent identified by religion or non-religion.

Salient Features: Social Position and Identity

Space and time were clearly critical factors shaping women's conceptualisations of their position and identity. Some women adopted the conventions of assuming their own class status from partners or parents. Class and colour were identifications more commonly used by white women than black.

Women in general appeared to resist identification as disabled women, perhaps because of negative connotations. Women may have assumed the identities they related to that have the highest visibility in public discourse or that they had the most to gain from. Social position and identity were informed by personal current experience, current roles and relationships, spatial context, surrounds and environment that is by *present time* in women's lives. For example, adoption of the identities *mother* or *friend* reflected current practices of mothering or friendship. In addition, *past times* were critical in shaping position and identity, that is, processes of urbanisation, industrialisation and immigration and memories of personal, familial and community past. For example, the history of immigration was uppermost in black women's self-identification, whether they were born in the U.K. or not. Countries of heritage were frequently mentioned by these women as a source of identity. Social position and identity were also clearly influenced by expectations of *future time*, for example, desires of future work. Women's reasons for entering higher education related to ideas about themselves and their households in the future. For example, some women who identified as middle class did so despite current low income and perhaps expressed a hope for the future as well as a memory in the past.

Some women used the opportunity to self identify as a space for critical reflection on their social position. They showed awareness of the ways in which both social identity and social position are spatially/temporally produced. For example, Grace recognised that her new sense of self has arisen from coming to college: "*I am a mature student from a working class background. I am female and white. Education has opened my mind to these issues. Previous to becoming a student I don't think I could have thought of these identities.*" Zandra was ambivalent about her social identification as an Afro/Caribbean [*sic*] woman. Although she valued her geographical heritage she felt that the use of the term represented social limitations in her life: "*The fact that I'm [an] Afro/Caribbean woman, born in the West Indies and brought up in England. I realise the differences between black and white i.e. culture conflict and the fact that I'm still known as Afro/Caribbean and my own experiences as a black working woman makes me realise that radical changes will take a very long time.*" Prabha, an overseas student, recognised that her specific spatial location and origins gave her a unique standpoint into wider social relations: "*The fact that I am female and I am black with different socialisation and culture. I think it has helped me to know who I am and where I am going. It has made me . . . see differences in society culturally and enabled me to respect other cultures as well as avoid cultural dominance.*" Rian saw higher education as a space that had legitimated an identity she had craved: "*Being . . . on a social science course has been a real revelation to me. It has given me loads of reasons as to why I feel the way I do about a lot of things and has also made me feel a lot less guilty for not choosing such a clearly defined 'womanly' role. I now realise that I am not alone in feeling and thinking the way I do about things and these feelings and thoughts aren't 'abnormal' or 'bad'!!"*

The lives of women students were highly complex. Women identified themselves in terms of official classifications, but also in relation to the sets of social, spatial and temporal arrangements they found themselves in. Space and time were critical in shaping social position and sense of self. Some women critically reflected on these issues in constructing new identities. They mediated the dominant spatial/temporal practices of classification with their own space and time.

Spatial/Temporal Practices of Paid Employment

Past Times of Paid Work[21]

Thirty-four students said they had been in paid work prior to studying. Forty-four different jobs were mentioned. I have chosen a classification of posts based on students' own terms of reference, because of the difficulties of using official classifications. In general most jobs mentioned fell into the official categories of semi-routine and routine occupations, although there were some exceptions to this. The preponderance of care and sales related occupations reflected gendered divisions in the labour market, career choice, prior experience and local employment opportunities.

Present Times and Paid Work[22]

Maureen gave an account of eight different jobs undertaken whilst a student, some simultaneously: "*Esso Petrol Station—cashier—21-hour/week—£3.75/hour, cleaner—7.5 hour/week—£3.80/hour . . . Young people's centre—12.5 hour/week—£6/hour . . . Young people's project—7 hours (as and when needed)—£6/hour . . . Worked as a delivery driver and cashier at Chinese take away. Did Klean Ezzy—i.e. delivered catalogue . . . and collected orders.*" The effective management of time and space was clearly crucial to sustaining multiple paid work commitments. The fragmentary and insecure nature of paid work experienced by women students, and the way their time was regulated at work is evident in Susan's description of the restructuring of her post as a care worker: "*They've knocked the hours. The full time week is no longer forty hours it's thirty-seven . . . they've decided they're going to look into cutting enhancements on weekends . . . Flat pay . . . and you work five days over seven. So that would mean because I work unsocial hours ... I would lose about £150 a month ... Union's fighting it. They're now talking about grading jobs."*

Forty women identified eighty-one paid jobs they had had whilst studying.[23] As above I have classified these through students' own terms of reference. Again, most jobs mentioned fell into the official categories of semi-routine and routine occupations, although there were some exceptions to this which related to the students' long term career choice (for example, in pursuing vocational qualifications in leisure and youth and community work).[24] Nearly a third of women had experience of care work whilst studying, followed by food and drink, domestic and cleaning work and sales and retail. Patterns of work also appeared to be age and accommodation related. Café, bar and shop work was generally the remit of younger women. Nearly half of the women living in halls had done sales and retail work, nearly a third cleaning and nearly a third café/bar work.

Women students were subject to a *time squeeze* that had implications for what they could do, where and when they could do it. As students their time for paid work was reduced but often their need to sell their time to obtain the means to live increased. This situation was exacerbated by cuts in state support for higher education.[25] Only six of the forty-six women did no paid work at all whilst studying. It might be expected that these students would be those with particularly heavy demands in relation to child care but in fact sixteen of seventeen mother students did paid work during the course, including six of the seven lone parents.[26]

The jobs mentioned by women included holiday work and work in major conurbations near to the town where the college was situated. However, the profile of jobs clearly reflected employment opportunities in the local community where the bulk of students lived. The college was situated in a small town in a picturesque rural environment. The town depended on tourism and there were many small cafés and bars that reflected this market. There were also a large number of private residential care and nursing homes in the area because of the availability of a substantial number of large Victorian houses considered appropriate for residential care purposes. In addition the college itself provided students with work as cleaners and in catering (in the college canteen and student union bar).

The shift in emphasis from work prior to college showed an increase in cleaning work and a slight decrease in clerical and administrative work.[27] This may have related to job availability in the community, and the fact that office work would coincide with the education timetable. The work that women students were involved in was clearly gendered. The vast majority of posts were in one way or another connected with the bodily needs of others. Care work involved assistance with bathing, toileting and meals.[28] In cafés and bars women were involved in the collective provision of meals and drinks. In cleaning jobs they removed human waste products and the debris of human activity.

Christine Cousins[29] argues that the development of industrial capitalism and process of urbanisation has led to *"functions provided by the family in the past, such as recreation, leisure, security and emotional needs [becoming] commodities, as too did the care of the young, the sick."* Such work is associated with women, because of horizontal and vertical gendered segmentation in the labour market. Claire Callender points out, *"Half of all women work in just three occupations: clerical and secretarial; personal and protective services, such as nursing, catering, cleaning, hairdressing; sales occupations."*[30] This pattern of work intensifies where women work part-time and in the case of these women students was intensified further. The world of work provided a particular niche for women students. The restructuring of paid work has facilitated this, for example, *"the promotion of flexible, efficient and competitive labour market"* since the 1970s[31] and the restructuring of welfare, which involves *"the promotion of a welfare market, tighter financial control, tighter management and privatisation."*[32] The consequence of these developments was that the jobs available for women students were usually at the periphery of organisations, associated with low pay and limited rights. For example, in relation to hourly rates of pay,[33] women on average earned between £4.36 per hour (production work) and £3.56 per hour (food and drink). Caring and cleaning posts paid on average £3.81 to £3.87 per hour.[34] Nineteen percent mentioned low pay and 16 percent unsociable hours. Sixteen percent also felt they had been actively discriminated against in paid work because they were students.

It was not possible to ascertain exactly how many hours that students worked on average per week. Patterns of student participation in employment varied in relation to the time of the academic year, stage of the course, holiday and religious periods. I have identified the average hours women said they worked per post, per week. I have excluded full-time work and work explicitly identified as holiday work. This was because of the difficulties in identifying whether work identified as full-time was holiday or term time work.[35] On average women worked thirteen hours per post, per week.[36] Lone parents worked slightly less (eleven hours). What is particularly striking from the figures were the hours worked per post by mother students who lived with a partner (fifteen hours). I had expected hours per post to decrease in relation to increased household labour. Although this association is evident in the case of lone parents, those mothers living with a partner worked the highest number of hours per post. This may have reflected the high costs of living in families and also support evidence that access to paid work varies in relation to whether women have partners who are in paid work.[37]

Paid Work and Studies

In order to begin to explore student perceptions of the links between paid work and studying I asked an open question about how they felt their choices and experiences of paid work had affected/been affected by being students. This was a difficult question, worded clumsily, because I did not want to imply one way causality, for example, by assuming that being a student has *caused* shifts in patterns of working. This would undervalue the active choice of becoming students, which is part of a complex change process. The purpose of the question was to try and capture some of the complex interconnections between paid work and studies. Thirty-two students responded to this question often mentioning more than one issue. Half the women felt that work opportunities had been limited by studying—*"Packing—was very low status, under paid and overworked . . . and the pay unsatisfactory!"* (Suraya)—but 16 percent felt they had wider paid work opportunities as a result of their studies: *"Started paid employment at 14 in catering. Been studying now for six years. Opened up opportunities I would never [have] otherwise encountered—tried new things etc."* (Katrina). Rachel summed up the ambiguous role of paid work in women students' lives: *"I have been able to move on in my work because of the course, but have to juggle my work life around the courses."*

Although most students emphasised their limited choices in relation to paid work, in fact the relationship between studying and paid work is more complex. A significant minority referred to widened opportunities in relation to paid work. On the one hand the requirements of studying (time) had a limiting effect on the type and condition of work open to students. On the other hand, being a student in higher education was perceived as affording wider opportunities, particularly in terms of developing the skills and experience for future work. The college introduced students to concrete settings for work experience through placement and other employer contacts. In the main however, students did not perceive paid work whilst studying as a bridge to better career opportunities but as a drain on their time, energy and resources for studying. Their work pattern prior to college, of low-grade routine and stressful work was perpetuated in most cases whilst they were students. The occupations available were often emotionally as well as physically draining as students, for example, some said they had to assist with death and dying in local care homes without sufficient training and support.

Future Times and Paid Work[38]

When asked what their long-term aims were in terms of paid work, forty-five women mentioned forty-eight jobs in total. As above I have chosen a classification of posts based on students' own terms of reference. Most of the occupations mentioned fell into the categories of higher professional and lower managerial and professional. Women students' aims were consistent with their educational choices and there was weighting to youth and community work, social work and health. Six of the seven lone parents and five of the seven black students wanted to work in community and/or youth work. The emphasis on interpersonal work reflected women's occupational and academic interests, module choices and gendered divisions in the labour market. Higher education was perceived as a route from routine work to more autonomy in work and better pay.

Salient Features: Paid Work

The dominant *spatial/temporal practices* of paid work gave rise to the following salient features for women students. Part-time, temporary work was the norm. Some women sustained more than one post simultaneously. Many worked unsocial hours. Most received low pay. Most worked in occupations directly related to bodily needs, for example caring, cleaning and food/drink. Some worked in a small pool of jobs directly associated with future careers, for example, community work.

Spatial/Temporal Practices: Housing and Household

Housing Tenure[39]

In relation to housing, women students were often in transition. Key transitions related to ageing, leaving the home of birth and establishing an independent home.[40] Other transitions related to relationship break-up, children, health and ageing parents. Choices around housing related to these transitions, to the amount of money available, to proximity to college and also to women's perceptions of safety. Forty-five students responded to a closed question about current housing. Over half rented their homes; one-fifth lived in the parental home and one-third were owner-occupiers. Six of the nine women who identified as middle class were owner-occupiers whereas only four of nineteen women identifying as working class owned their own homes. Of those renting, 60 percent lived in halls of residence. These were mostly younger women, but also a few older women moved into halls particularly after relationship breakdowns. No women of South Asian heritages in the sample lived in halls of residence. This may be due to the fact that their families lived within travelling distance of the college, although issues of safety may also have informed the decision. Eighteen percent rented from the private sector, 18 percent from housing associations and only 2 percent from the council. Six of the seven lone parents lived in rented accommodation.

State restructuring of housing provision since the 1970s shaped women's choices in relation to housing. There had been a massive reduction of social housing provided directly by the state and this was reflected in the small percentage of women students living in local authority accommodation. The growth of the housing association sector and the growth in owner occupation were both reflected in women students' patterns of residence. High rents were characteristic in all the rented sectors as a consequence of the deregulation of rents and the abolition of fair rents.[41]

Housing Conditions[42]

Forty-five students responded to a closed question about housing conditions.

Those who perceived their housing conditions as least adequate were women who lived in halls of residence. Over half of those in halls said that their conditions were overcrowded and nearly three-quarters referred to deterioration: *"Quite a big room with two beds. Most of the furniture is broken and battered. Walls affected by damp and peeling paint. Big bubble in carpet. Floor slants quite badly"* (Olivia). Students also referred to the accumulation of waste and the intensive nature of daily living: *"Have to walk past rubbish to bath . . . Have to borrow Hoover. Have to do it at a certain time. Washing . . . 3 machines for 40 . . . Up and down till one's free . . . Things breaking in maintenance book . . . Only one shower works—only 2 for 40 people"* (Rose). The remainder of women students generally perceived their housing as mainly good or good throughout, in particular those women who identified themselves as middle class. Lone parents who rented their homes particularly mentioned heating problems (see below).

Housing and Study Space

When asked where they studied in their accommodation, forty-four students responded, sometimes raising more than one issue. Only five women had one fixed place to study that was not used for any other activity and not shared with other people. Nearly half studied in two or more places in the home. In over three-quarters of cases the places of study were also used for other activities such as cooking, childcare and sleeping. Where women identified two or more places for study, sometimes they said it was because of the "squeeze" on household space and sometimes a more positive choice. For example, many mother students had no choice but to combine studying with watching over children, but some younger women living in the parental home said they preferred studying with their families around them as they felt less isolated (see chapter 8).

When asked to describe significant tensions over quiet space to study nearly one-third of women referred to relationships and caring. The impact was higher in the case of those who lived with partners and/or children—"*I study in the spare bedroom, tensions are limited to my partner wanting to play on the computer situated in this room. If this is the case I move downstairs and study in the lounge*" (Grace)—but also affected women in halls and in the parental home: "*Own room—share with younger sister. Kitchen—Can't. Mum studies there. Living room—TV distracting, end up helping with sister's homework*" (Janice). Women also mentioned noise as interrupting studies. Noise was the dominant factor affecting women in halls—"*It's best to study in the library as it's virtually impossible to even think straight in halls. No matter what time of day or night, it's noisy—this does sometimes have its benefits though!*" (Geni)—and "*Most tensions arise from lack of privacy; noise and living with forty other people who don't all sleep, wake up and have the same time to study*" (Sonya).

The need for warmth particularly affected lone parents (four of the seven), all of whom rented their homes: "*Difficult to study with only one room downstairs and three children. Can use bedroom but not very warm*" (Michelle). There was also reference to the need for adequate furniture and equipment. The ability to create space to study related closely to both the physical and emotional space available to women and the other demands on their time from relationships, household responsibilities, paid work and community/leisure.

Household[43]

Forty percent lived with their own children (half of these were lone parents) and 24 percent with a partner.[44] The rest either lived in halls of residence or with their own parents (see "Housing Tenure," above). A very small number said they lived with friends in rented accommodation other than halls (4). When asked an open question about the ways their households had changed since becoming students, women referred to a variety of life transitions that they perceived to be significant. Of the nineteen women responding to this question, a quarter had been affected by separation/relationship breakdown and a quarter had moved out of the parental home. A fifth had moved house and in a fifth of cases their own children or siblings had left home. Other transitions included moving in with a new partner, children starting school, having a baby and bereavement. Some transitions directly related to studying, for example, where someone chose to move into halls of residence. Others may have been unconnected, such as bereavement. Yet others were subtly interwoven with choices about studying, for example, separation may have arisen from the choice to study or the choice to study may have been an outcome of drawing apart from a significant other.

Housework

Negotiations over housework were essential to creating sufficient study time[45]: *"[I do] most housework as in cleaning [husband does cooking when home]. He will clean but not of a very high standard. Joint in looking after kids, but I take most responsibility for clothes, presents [Christmas/Easter], babysitters, doctors appointments etc."* (Christine). It was not possible to quantify the number of housework commitments women had because of the simultaneity of many housework activities (see chapter 6), but it was possible to identify whether housework was shared and to what degree. Thirty-nine women responded to an open question about housework. Lone parents and women in halls of residence said they did all their own housework (with exceptions). Women with partners said they either shared the housework equally[46] or did most or all of it. Only one of these women said she did no housework but that this was a source of much tension. Women who lived in the parental home said they contributed to the housework, with one exception.

Childcare

It was not possible to quantify the number of childcare commitments women had because of the variety of responsibilities they held, including caring for their own children, siblings, nieces and nephews and so forth. However, it was possible to identify whether childcare was shared and to what degree. Sixteen women responded. Lone parents said they did all the childcare with occasional help. Women in partnerships said they either shared childcare equally or did most or all of it.[47] Only one mother said she did no childcare but that this again was a source of much tension. Women living in the parental home also contributed significantly to childcare. Mothers referred to getting help from their own mothers, a father, siblings and partners. They also paid for childcare.

Household and Domestic Commitments and Studies

Thirty women responded to an open question about tensions arising from household and domestic commitments in relation to the need to study. Some raised several issues. The major tensions around studying included housework (43 percent): *"I get frustrated, angry, fed up, depressed and upset . . . not getting the support . . . if [housework] were shared I could do college. If I do housework while husband at home he gets annoyed as says we need to spend time together"* (Maureen). Emotional and relationship issues were raised by 27 percent of women, two-thirds of whom had partners and children: *"Husband's jealous of my studying. Experience of awkwardness, moaning about the house; books everywhere; late nights; sex life because I can't be bothered; constantly on the phone to fellow students in times of stress; hogging the computer"* (Rowena). Mainly lone parents mentioned childcare: *"My daughter spends more time at my mum's than at home—She now wants to live there as she receives more attention than what I can offer at the moment. Results in feeling torn between being a good parent and being a committed student"* (Katrina). Women in halls mentioned lack of private and quiet space to study, stressing this even though they had their own rooms and fewer domestic responsibilities than other women. One-third of those living in halls mentioned noise (see "Housing and Study Space," above). Mothers mentioned finding time to study and five women mentioned money worries.

The profile of household tension in relation to studies reflected the three classic spheres of women's household labour, that is, housework, emotion work and childcare. Women discussed the ways they were expected to give space and time to others within the household through household labour and relationships. They were also expected to give space and time so that others could sell their time in the labour market. This involved negotiations in the household about the value of time spent by certain people (sisters, brothers, mothers, fathers, daughters and partners) in certain activities. Questions arose as to who should sell their time in the labour market, who should give their time to others in the household, the value of paid work versus education and the value of housework versus paid work. Time spent in paid work is increasingly valued most highly even by women.[48] There is increasing ambiguity in public discourse regarding the value of time spent in nurturing and responsibility. This has been highlighted in debates about *good mothering*, the decline in men's work, and the increase in women's paid work. A focus of a lot of this debate has been lone parents and paid work.[49] However, whether or not women lived in traditional households, as lone parents or in same sex households, their lives were influenced by these discourses.[50] Women who wanted to study in higher education had to both rationalise and negotiate their space and time to do so through this hierarchy of values. The concept of higher education shifted in meaning in the process. Sometimes it was conceptualised as work time, sometimes as time for the household. It could be conceptualised as time for others, as selfish time and as time for self. These issues are followed up in chapters 8 and 10.

Women living in collective housing (halls) may have appeared to be free from such household negotiations. They may have negotiated for financial support but at a distance from their parents. However, they were caught up in the flux of friendships, relationships, housework and paid work and negotiating this terrain could be just as intense. Women in halls wrote of boyfriends demanding their emotional time and of the chaos of living with large numbers of other students. For a few older women, halls were perceived as a space for them to study (after or during relationship break-up) however *others' space and time* invaded their own in halls, particularly as noise and constant interruptions.

Salient Features: Housing and Household

The salient features of the housing and household arrangements of women students were as follows. Some younger students stayed in the parental home. This was predominantly because of the need to self finance, but may also have been affected by their perceptions of safety in other forms of housing, geographical proximity and perhaps because relatives may have needed personal assistance. Most mature students in partnerships were in owner occupation. Both partners tended to be in paid work. Those renting (other than in halls) tended to do so from the private or independent sector. Most lone parents rented and raised issues about affordable heating. Most young students lived collectively in halls. These were old buildings in a state of considerable deterioration. There were certain collective rules, but more generally the accommodation was far less regulated than other collective or institutional provision (for example, care homes). Women discussed the intensity of living, noise and relationships. The women in this sample in halls were all white. Negotiations within households related to the value of time spent on certain activities (housework, childcare, paid work, leisure and education).

Spatial/Temporal Practices: Leisure/Community

Leisure/Community Activities[51]

This sphere of activity is highly complex and interconnected with the spheres of paid work, education and household. The public provision of leisure has features that do not necessarily accord with people's actual leisure experiences. For example, leisure facilities have historically been class, gender and racially segregated and may be sites for the leisure of some groups and not others.[52] It was difficult to quantify women students' leisure experiences. To measure use of formal leisure spaces is unsatisfactory because one might not actually experience leisure there. For example, tensions persisted in leisure spaces such as the student union bar, where both black women and a lesbian woman said that they had felt uncomfortable (see chapter 9). There are complex meanings attached to leisure and problems associated with women conceptualising time for self. This complexity has been associated with unequal gender relations as well as other social divisions in relation to both the time available for leisure and structured inequalities in the leisure market.[53] It was possible however to identify those activities which women perceived as time for leisure and community activities (such as schools, nurseries and community groups) and to identify their pattern of involvement. Women were asked what community, leisure and social activities they were currently involved in. Forty-six women responded detailing sixty-nine activities. Responses have been classified according to the analysis discussed by Karla Henderson et al.[54] together with women's own terms of reference.

The range of responses was wide and included such comments as, *"On management committee of . . . nursery"* (Katrina), to, *"A bath is a luxury—especially if the kids don't knock for the toilet every 10 minutes"* (Christine). However, most written responses fell into traditional conceptualisations of leisure. Home-based activities were least mentioned in the questionnaire although a major source of leisure for women. Women in halls mentioned social activities and friends to a much greater degree than those who lived in the parental home. Women in their forties also mentioned high levels of social activity around friendships. Five of the seven black women students identified a religious centre as a source of leisure, social and community activity, a much higher proportion than amongst white students.

When asked whether there had been a change in their activities since becoming students, thirty-nine women responded. Fifty-one percent of women said there had been no change in their leisure participation, 33 percent a decrease and 15 percent an increase. However, the proportions shifted when mother status was taken into account. In the case of mother students over half reported a decrease in leisure time and only one an increase. In the case of women with no dependent children, particularly those living in halls, more reported an increase in leisure than a decrease: *"Before I came to college my life seemed to be much more structured. Now it seems that things happen with more spontaneity and things aren't so routine"* (Geni). One-quarter of those identifying as working class said they had no leisure, whereas none identifying as middle class said this. This may have been because working class women had less free time, but evidence of their engagement in paid work does not support this as women identifying as middle class were more likely to be in paid work. It may be because of an increased burden of gendered work (unpaid) on working class women, who are less able to purchase services such as cleaning and childcare. It may also support Paul Corrigan's[55] conjecture that *free time* is not necessarily associated with leisure for low income and unemployed groups.

A complex relationship between leisure time and studying emerged. For single students and those without children a substantial number took new opportunities to develop social and leisure activities through the contacts and facilities arising from their entry to higher education. For women with high levels of paid and unpaid work, time for leisure appeared to be significantly reduced. Becoming a student appeared to involve an active transformation of the concepts and practice of leisure. This is further discussed in chapter 8.

Friendships and Relationships

Leisure for women students was a sphere for the building and cementing of friendships and relationships and also for the furtherance of household obligation.

When asked in what significant ways women felt studying had affected/been affected by their friendships and relationships, a complex pattern emerged.[56] Forty-four women responded to this question. The wording, although clumsy, was again intended to avoid implying one way causality, for example by suggesting that becoming a student had led to changes in relationships when it may well have been an outcome of such changes. I have clustered responses into three categories although this conceals a wide range of variation. These categories are *feeling supported, difficult transition* (which included, for example, regret at having moved away from friends or family) and *feeling lack of support*. Fifty-two percent of women said they gained support from friends, 41 percent from family members and 5 from partners/boyfriends. Thirty-nine percent said the transition had been difficult in relation to friends in particular and six women felt unsupported by friends.

Friends were clearly significant in all three areas, being identified as a source of support, as well as a source of difficult transitions and lack of support. The complex roles of friends are exemplified by Geni: *"We encourage each other to put the work off until the very end. Also by being here I've met several people who are very special to me and are now a large . . . part of my life"*; the ambiguity of friendship, by Sonya, *"The number of friends and social life has increased however tensions have arisen from living in halls. College pressures have meant that arguments arise due to stress and lack of personal space, privacy."* The concept of friend itself implies a supportive relationship[57] but is also used to describe close relationships that are distinct from family or kin. Shifting friendships marked becoming a student for many women: *"I feel that friends relate to me differently since I've come back to study"* (Kelly). When mother students were identified separately, family became a more significant source of support than for non-mother students. In relation to partners and "boyfriends," fewer than half of those women with partners identified them as a source of support. Five of the six referring to boyfriends perceived the relationship as a source of difficulty in relation to studies: *"Sometimes I feel that my boyfriend wants me at home every weekend, leaving me with no time to really study"* (Sally).

The dominant temporal rationale regarding leisure is that it is time for self, and therefore in the hierarchy of social values accorded to time and space, leisure has lower status and lower priority. Rosemary Deem[58] discusses the associations between paid work, leisure and the concept of entitlement. Leisure is sometimes conceptualised as a right emerging from paid work. In fact leisure is often time for others, for relationship building, for the extension of household activities and relates to struggles over resources for the household.[59] The boundary between community and leisure is blurred, certainly so when there is a need to justify time spent in these ways. As with women student's experience of higher education, leisure time was ambiguous, complex and contradictory and this came through in their accounts of their negotiations with friends and family. For example, young women in halls generally said they had increased leisure time. They may in fact have been describing involvement in certain activities designated as *not work* and as time for self. In reality these activities may have included difficult negotiations with boyfriends over study time. Such negotiations within settled partnerships would be less likely to be identified as leisure and more likely to be seen as household labour. On the other hand, mother students may have said their leisure time had decreased when they became students, because their formal involvement in certain activities officially designated *leisure* had reduced. However, for these women higher education itself may have been a source of leisure (being their only opportunity for time for self). These issues are followed up in chapter 8.

Salient Features: Leisure/Community

Women generally conformed to traditional conceptualisations of leisure in the questionnaire. This highlights the limits of this research tool and much more is revealed through interviews (see chapter 8). Women focused on formal leisure activities, friendship and voluntary activities. However, smaller numbers referred to leisure as an experience, for example, as time alone. For young women in halls and women in their forties the leisure focus was mainly related to relationship building and friendship. Both groups said they had high levels of social activity in relation to friendship. Between those ages women appeared to be more family centred, engaging in leisure activities for and with other family members. Women of both African/Caribbean and South Asian heritage identified a religious centre as a leisure resource. This perhaps reflected historical patterns of segregation where community based activities (including black-led religion) provided a vital role in providing safe non-racist space. Class identification was significant in conceptualising leisure, as free time did not necessarily correlate with leisure time for women identifying as working class.

Chapter 7

Conclusion

In the above discussion I have constructed *snapshots* of women students' everyday experience. This concept gives visibility to the physical movement of women: *"Space is divided up into designated (signified, specialised) areas and into areas that are prohibited (to one group or another). It is further subdivided into spaces for work and spaces for leisure and into daytime and night time spaces."*[60] For the women students involved in this research, there was no single space of higher education. Space and time for higher education had to be carved from, and integrated into other space and time. The dominant *spatial/temporal practices* of home, paid work and leisure shaped the lives of women students differently in relation to their social position and stage in the life cycle.

As Henri Lefebvre[61] argues, the spaces of paid work are not merely physically constructed but arise from acts of productive labour, from the practices of the state, from divisions of labour and from market and property relationships. Women students were, in the main, facing reductions in weekly income coupled with increased costs arising from studying. Household negotiations in relation to money were often intensified. The niche in the labour market open to these women because of their skills, qualifications and availability (time), were routine and semi-routine occupations characterised by low pay, lack of autonomy, a competitive work ethos and pressure to *produce*. Some women worked in two or more jobs simultaneously. Although there were patterns of difference in relation to age and residence, women students worked unsocial hours and in occupations heavily associated with the body, whether in care, sales, cleaning or food and drink. Gendered divisions of labour were reflected in women students' past and present experiences and their future expectations of paid work. Paid work, for many, involved an intensification of gender as their former home-based experiences were duplicated in institutional settings. The contemporary practices of paid work *"crush time, by reducing difference to repetition and circularity."*[62] In this case, the home-based tasks of caring, cleaning and/or preparing and serving food became repetitive and routine in the paid workplace. Women students hoped to move up the class hierarchy and gain more autonomy through higher education, but there was no significant shift in the gender characteristics of the future work they desired. It was still heavily associated with social care. However, a few women found new opportunities in the labour market whilst they were students, which were associated with their future career ambitions and were established through their connections with the college.

In relation to housing, a critical issue was the shrinking role of the state. This affected access to both housing and higher education. The deregulation of the housing market had led to the growth of owner occupation and increase in private rented sectors. Both these factors had cost implications for students. A few women found it hard to find warm affordable housing, in particular lone parents. The reduction in financial support to women students increased material dependencies on others, both parents and partners. These factors impacted upon the attempt to create study space and time. *Home* is indelibly connected and carved from the wider spheres of community and work. The period of their higher education studies, was, for many women students, also a period of household transition. There were normative expectations that young women would move from their household of origin at some point. Some older women were also in transition; for example, forming new households after relationship break up. *Home* was more than a dwelling place: "*The fact remains that a homebuyer buys a daily schedule and that this constitutes part of the value of the space acquired.*"[63] Lefebvre refers to the geographical distance between home and paid work, but clear considerations for the women in this research were the distance to schools, the shops and the college, in addition to childcare and housework schedules. Of those with their own children, virtually all did *at least* half, and frequently substantially more of the childcare and housework. The practice of collective living for younger students had direct implications regarding space and time for studies. Proximity to college did not necessarily facilitate academic work. Although young women mentioned the many positive aspects of living in halls, in relation to academic studies they referred to the negative effects of collective living, in particular the noise, the dirt and the poor health they experienced.

Leisure and social activities that women students referred to were generally associated with places external to the home. Dominant practices of leisure were difficult to quantify because of the ambiguity of concepts of leisure: "*Typically, the identification of sex and sexuality, of pleasure and physical gratification with 'leisure' occurs in places specially designated for the purpose.*"[64] Women students' concepts of leisure varied according to their social position. Responsibilities to others in the home meant that for many mother students, leisure time was conceptualised as time for the family. Friendship and supportive relationships were an essential ingredient of the struggle to find space and time to study (see chapter 9) and shifting friendships marked students' leisure time. This research instrument reveals little about women's intentions in their leisure choices but it does reveal patterns of leisure practice and particular pathways for some groups of women which may reflect both individual and collective intentions and resistance to forms of oppression.[65]

The dominant *spatial/temporal practices* of everyday life created separate and distinct spheres of activity for women students and particular routes through space and time. Differences in residence, household, age, colour, geographical heritage, health and ability had complex implications for the availability of space and time to study and the ability to carve out the necessary space and time. The following chapters move on to explore the meanings associated with different space and time and the ways that women carve out space and time in order to study.

Notes

1. Henri Lefebvre, *The Production of Space* (Oxford: Blackwell, 1991), 38.
2. Lefebvre, *Production, Space*, 38.
3. The rationale for selecting this sample of women students is discussed in chapter 6.
4. Patricia Hewitt, *About Time: The Revolution in Work and Family Life* (I. P. P. R., London: Rivers Oram Press, 1993).
5. Nira Yuval Davis, *Gender and Nation* (London: Sage, 1997), 131.
6. Closed questions were asked in relation to age, household status, paid work, housing and so forth. In addition, the dominant concepts of class, ethnicity, gender, sexuality and disability were drawn on in both questionnaires and interviews.
7. Many of the women used dominant categorisations in their self-descriptions. Women may well have been influenced by previous monitoring exercises, by their academic studies and by the format of the questionnaire itself.
8. Appendix 2a.
9. Questions were asked about the route whereby women had come into higher education and their qualifications on entry. Some students answered these in terms of the institution they had come from and some in terms of qualifications they had previously pursued. Most students gave both sets of detail. In terms of the institution, thirty-six students responded. In terms of qualifications on entry, thirty students responded.
10. Advanced (A) levels.
11. The question was deliberately worded in such a way as to encourage rounded responses from students, following Joan Parr, "Theoretical Voices and Women's Own Voices: The Stories of Mature Women Students," in *Feminist Dilemmas in Qualitative Research: Public Knowledge and Private Lives*, edited by Jane Ribbens and Rosalind Edwards (London: Sage, 1998).
12. Beverley Skeggs, *Formations of Class and Gender: Becoming Respectable* (London: Sage, 1997).
13. bell hooks, *Talking Back: Thinking Feminist—Thinking Black* (London: Sheba, 1989); Elizabeth de Silva, *Good Enough Mothering? Feminist Perspectives on Lone Motherhood* (London: Routledge, 1996).
14. Paul Gilroy, *Small Acts: Thoughts on the Politics of Black Culture* (London, Serpent's Tail, 1993).
15. Gilroy, *Small Acts*, 14.

16. African/Caribbean women were more likely to say they had particular career aims and financial autonomy. This may be because of the overlap with their status as lone parents.

17. Four of these women relied on parental contributions and two were mature students.

18. Only three women felt they had equal control of income with another. In all cases this was with a partner.

19. Jan Pahl, *Money and Marriage* (Basingstoke: Macmillan Education, 1989).

20. Thirty-three women responded to an open question about the ways they perceived their income to have changed since becoming students.

21. Appendix 2b.

22. Appendix 2c.

23. In order to identify the range and type of paid work, I asked the respondents to describe a current or recent paid post including hours and rate of pay, and also to list other posts they had had whilst studying.

24. Although a minority of posts mentioned might be officially categorised as professional, the temporary and insecure nature of the work, and the expectations of the students involved would not necessarily support this categorisation.

25. Whereas the state formerly paid for students' space and time for higher education, they were now expected to be mainly self-financing.

26. Three of the six women who did no paid work were from households involved in giving personal assistance to a disabled or ill person.

27. There were no significant differences in the type of work done by those identifying as middle or working class.

28. Mothers with partners and women living in the parental home were more likely to have engaged in care work.

29. Christine Cousins, *Controlling Social Welfare: A Sociology of State Welfare Work and Organisation* (Sussex: Wheatsheaf Books, 1987), 66.

30. Claire Callender, "Women and Employment," in *Women and Social Policy: An Introduction*, edited by Christine Hallett (London: Prentice Hall, Harvester Wheatsheaf, 1996), 38.

31. Callender, "Women, Employment," 41.

32. Cousins, *Social Welfare*, 142–49.

33. In order to work out hourly rates of pay, I selected only the most typical occupations held by women students and arrived at an average where either hourly rate was directly mentioned or easily arrived at through dividing hours worked by weekly pay. Figures are based on gross income. I excluded full-time work because of the difficulty of establishing the hourly rate. No significant differences between women identifying as middle class, working class or by colour were evident.

34. Appendix 2d.

35. Appendix 2e.

36. The figures above may well *underestimate* the actual hours worked per week by women students during term time. For example, when asked to identify a current or recent post nearly a quarter of women mentioned two or more posts that may well

have been done simultaneously. It is also common knowledge that some students continue in full-time work throughout term time and this has been excluded from the figures.

37. Report of the Commission on Social Justice, *Social Justice: Strategies for National Renewal* (London: Vintage, 1994), 184.

38. Appendix 2f.

39. Appendices 2g and 2h.

40. Claire Wallace, "Between the State and the Family: Young People in Transition," *Youth and Policy* 25 (1988): 25–36.

41. Housing Act (U.K.), 1988.

42. Appendix 2i.

43. Appendix 2j.

44. Closed questions were asked in relation to household composition. Forty-seven responses are included because one woman identified herself as part of two households, having moved into halls to study.

45. Jo Van Every, "Understanding Gendered Inequality: Reconceptualising Housework," *Women's Studies International Forum* 20, no. 3 (1997), 411–20. She argues that such negotiations in fact construct gender and that the nature of such negotiations is perhaps more significant than the tasks actually carried out, in identifying tensions in the household.

46. A quarter of women said they shared housework equally. Nearly all lived with a male partner.

47. A quarter said they shared childcare equally. Nearly all the women in partnerships either shared childcare equally or did most/all of it with occasional help.

48. Karen Davies, *Women and Time: The Weaving of the Strands of Everyday Life* (Aldershot: Avebury, 1990).

49. Elizabeth de Silva, *Good Enough Mothering? Feminist Perspectives on Lone Motherhood* (London: Routledge, 1996).

50. See Sarah Oerton, "Queer Housewives? Some Problems in Theorising the Division of Domestic Labour in Lesbian and Gay Households," *Women's Studies International Forum* 20, no. 3 (1997): 421–30. She argues that even lesbian partners' domestic labour arrangements are gendered.

51. Appendix 2k.

52. Eileen Green, Sandra Hebron and Diana Woodward, "Women, Leisure and Social Control," in *Women, Violence and Social Control*, edited by Jalna Hanmer and Mary Maynard (London: Macmillan, 1987).

53. Karla A. Henderson, M. Deborah Bialeschki, Susan M. Shaw and Valerie J. Freyysinger, *Both Gains and Gaps: Feminist Perspectives on Women's Leisure* (State College, PA: Venture Publishing, 1996).

54. Henderson, Bialeschki, et al. *Gains, Gaps*, 172–94.

55. Paul Corrigan, "The Trouble with Being Unemployed Is That You Never Get a Day Off," in *Freedom and Constraint: The Paradoxes of Leisure: Ten Years of the L.S.A.*, edited by Fred Coalter (London: Routledge, 1982/1989).

56. I have conflated references to student friends with friends in general because women were not consistent in making this distinction. I have also conflated the identification of particular familial and sexual relationships for the same reason.

57. Eileen Green, "Women Doing Friendship: An Analysis of Women's Leisure as a Site of Identity Construction, Empowerment and Resistance," *Leisure Studies* 17 (1998): 171–85, 179. She argues that friendship may be conceptualised as a *"collaborative tool for exploring the world."*

58. Rosemary Deem, "Feminism and Leisure Studies," in *Sociology of Leisure: A Reader*, edited by Charles Critcher, Pete Bramham and Alan Tomlinson (London: E. and F. N. Spon, 1995).

59. Nancy Naples, *Grassroots Warriors: Activist Mothering, Community Work and the War on Poverty* (New York: Routledge, 1998). She uses the concept of *activist mothering* to describe women's community work participation in the U.S.

60. Lefebvre, *Production, Space*, 319–20.

61. Lefebvre, *Production, Space*, 191.

62. Lefebvre, *Production, Space*, 18.

63. Lefebvre, *Production, Space*, 339.

64. Lefebvre, *Production, Space*, 310.

65. Susan Shaw, "Conceptualising Resistance: Women's Leisure as Political Practice," in *Journal of Leisure Research* 33, no. 2 (2001): 186–201.

8

Spatial and Temporal Representations: Guidelines for Action

"What emotional paradox are we apparently trying to resolve in order to live the life we want to live? By what dialectical interaction between feeling, rules and context is this emotional paradox produced?"[1]

Introduction

THE AIM OF THIS CHAPTER is to discuss the research findings drawing on the concept of *spatial/temporal representations*. The main source of analysis is interviews carried out with seventeen women students (see chapter 6 for the criteria used to select students for interviews). Material from questionnaires is also drawn on where relevant and therefore more than seventeen "voices" inform this chapter. The chapter explores what women say about the different spaces they have experienced whilst students, including home, paid work and leisure/community. The discussion gives insight into the context within which women begin to create space and time to study and their potential to transform space. The space of higher education is discussed in chapter 9. The primary research question being addressed here is: "*What are the dominant spatial/temporal representations shaping women's everyday experiences whilst studying?*"

In interviews, women gave different but related accounts of the places they experienced whilst students. In pulling these accounts together, particular *landscapes* emerge. I use this concept to organise women's accounts of their experience. I construct *landscapes* related to the spheres of paid work, housing and household and leisure/community, and in relation to each sphere focus on the physical, social and emotional aspects, as outlined in chapter 3. At the end of each section I have identified key *spatial/temporal representations* drawn from women's accounts. I also identify areas of *emotional dissonance*. They may be areas associated with emotional unease, material insecurity or low social status. For example, conceptualisations of home as a safe haven create dissonance for women who face violence in the home.

Spatial/Temporal Representations of Paid Work

The Physical Landscape of Paid Work

The *spatial/temporal practices* of paid work are informed by international and local networks of capitalism. The physical landscape of paid work for women students also extended beyond national boundaries. Suraya and Rashida, for example, both in their early twenties, lived with their parents whilst students. They both referred to their mothers' working lives in South Asia. This influenced their perceptions of work. Both were strongly supported in their pursuit of higher education by their mothers, whose education had been curtailed through work: "Yes, well, she started off reading Arabic and stuff and . . . but because she was the only daughter and her mum became ill . . . She actually looked after her until she died and she had other responsibilities and she was always talking about you know so much work to do, work in the fields" (Suraya). Suraya's mother's experience of work was local, family based and extremely tiring. The duress involved formed a benchmark for the way Suraya conceptualised *real work* (see below) as associated with pain and discomfort. She described shift work doing packing in a factory and the ways that personal time for sleep was fitted into the dominant clock-led routines of the shift system: "Two in the afternoon to ten in the evening . . . We got half an hour break for dinner and two ten minute breaks, but that wasn't enough, it really wasn't. Especially the hours it was like sleeping during the day . . . this type of thing, going home, went home about eleven, twelve, get to bed about twelve, sleeping till about six, seven, eight, getting up again and going back to work."

A key aspect of the physical landscape of paid work for women students, was work in cafés, factories, care homes and retail outlets. This work was rigidly periodic, was usually outside the home and was often situated in medium or large institutions. Women involved in social care work painted vivid pictures of the routines and regimes of collective care. They talked of regimes developed to manage the bodily needs of residents and the impact of the physical environment. Two typical work routines demonstrate the physical landscape of care work in institutional settings. Rachel described a shift as a nursing assistant on a psychiatric ward: *"It depends which clients you have. Someone on level one, which needs to have someone in touching distance of them all the time, they might be suicidal . . . they're not allowed out of your eyesight. Someone that's very unwell . . . end up restraining them . . . and then others that might want to go out, because we have some rehab. Clients from prison . . . it varied all the time."* Zandra described a shift in a residential establishment for people diagnosed as having Alzheimer's disease: *"A lot of them rather than being in wheel chair . . . they walk around a lot, they're free . . . believe that they're still very young . . . so they're very brisky up and down the place . . . where obviously they don't know what's going on . . . when it comes to toilets and things they would do whatever anywhere in the home. . . You do have to clean up. You have to be patient."* The physical landscape of paid work in residential settings was both brutally intimate and impersonal. The buildings in which the work was carried out shaped routines within them. Anthony Giddens uses the concept of the *longue durée of institutions* to give visibility to the time of social institutions.[2] This concept reveals the ways, for example, past ideas about hygiene, health and sexuality shape contemporary institutional practice. Much of the work that women described concerned personally assisting groups of people whilst maintaining boundaries within institutions and managing particular uses of space. Residents sometimes resisted this. For example, Irene discussed the bath routines in the home she worked in. So many residents were to be bathed on a particular day. Deviation from this routine was frowned on by both managers and other staff: *"I thought that that was a bit strange but then I thought that there was no other way of doing it. You can't give the same environment that you'd have at home . . . it is a business . . . I thought about it . . . But what other way could you do it? Although these people have rights to be able to say when they want a bath . . . you can't always do it when they say"* (Irene). In the past such regimes may have been rationalised through the idea of collective care, but the escalation of community care policies and the associated privatisation of parts of the care sector, facilitates a business oriented rationale for such practices.[3] Behaviour which students had found natural outside the institution was frowned on within it: *"A particular lady I used to talk to. I used to do quite a lot of things for her really . . . she was so interesting. I got told that you can't spend all your time with one individual . . . When I used to go on early shift . . . I used to make sure that they were all washed, clean, nice clean clothes. I used to ask them what they wanted and because I wasn't quick enough . . . You get looked at as if you're a bit of a swotty"* (Irene). Zandra used the active concept doing care to describe her paid work rather than the passive concept of being a carer. This gave visibility to her as actively engaging in work assisting others over time. Her words were far more expressive of the myriad of daily intimacies involved and the lack of privacy experienced by residents and workers alike. Zandra also worked in a small residential unit for older people on a night shift. The smaller unit generated a wider range of work demands than the larger institution she worked in (see above): *"At this particular place, there's two of us on, on a night . . . one of the first thing [sic] you actually do is go round and give the medication out. Then these residents are assisted to the bedroom where you'd wash them, dress them, get them ready for bed, put them into bed . . . you would actually start the cleaning task . . . you would Hoover up . . . you would polish. You'd set the dining table . . . help the cook by peeling the veg."*

In domiciliary care settings, the regulation of space, time and tasks was equally if not more frustrating. Susan, who worked as a home care worker for a social services department, spoke of working with a woman who was very poorly with terminal cancer. Susan attempted to respect the rights of her client to control her own space but management failed to recognise that extra time was therefore needed for personal assistance to be carried out effectively:

> You are paid by the hour . . . This lady, she would like get up, get washed and dressed and then she'd go sit in a chair and you'd make her breakfast. Well that sounds really easy, but then she would become really poorly and then she's really really slow and she would drag herself there and she's determined and so particular and everything had to be just so . . . Not just how she was, how things around her were. And you just couldn't get away. So, I mean I sort of like said [to management] "It's taking a lot longer and I'm not getting to clients till such a time," and they'd just say, "That's how it will have to be."

The management of physical space and time in the retail sector was also a central feature of the work regime. Rose described the hourly rota at Top Shop. This particular regime was experienced by her as *"worker friendly,"* but it also clearly emerged from the process of managing a large-scale retail outlet with multiple functions, to maximise efficiency and profit. The regime was one where potential customers moved through the building looking at, trying on and buying clothes, returning goods and sometimes shoplifting. Their movement was reflected in the periodic movement of staff through the building to carry out different tasks: *"Top Shop rota. One hour fitting room. One hour till. One hour at front keeping an eye out . . . One hour at Top Man. Selling personal account meant bonus."*

When she worked in a warehouse Suraya mentioned the pressures from management to increase productivity. She spoke of the pressures from the "main boss" in the warehouse where she did packing. He was principally concerned to maximise the regulation of workers' time: *"It was a small group . . . different shift two to ten and so on our shift there was probably about twenty of us. The supervisors and that were OK, it was just when the main boss came. He was a bit, you know, 'you should be doing so many . . .' and stuff."* The use of space in paid work settings sometimes constructed and reinforced boundaries amongst both clients and staff. Suraya mentioned the strict "racial" demarcation in her work with older people in a community centre: *"I worked at an elderly day care centre . . . worked with Asian elderly and white in two separate rooms."*

Cleaning and domestic work also involved rigid regimes, this time in relation to the order and intensity of cleaning tasks. Much of the cleaning work was directly concerned with dealing with personal waste in institutional settings, including residential homes and the college where they studied. Ruth spoke of the ways times external to paid work (studying and housework) limited the type of work she was able to do and drove her to cleaning: *"I do feel that they [work opportunities] have been limited due to the time factors, i.e. certain types of jobs are only available during the day. Cleaning after office hours. Washing dishes in the evening."* As with care work, the bodily times of other people shaped cleaning work. But in this case it was carried out in the absence of those people it benefited and in physically empty spaces.

Familial and community memory shaped women students' concepts of physical labour and informed women's relationship to the physical landscape of paid work. In addition the physical landscape of home overlapped with the landscape of paid work as practices of home-based caring and cleaning were translated into paid work settings. The key physical representations of paid work were fourfold. Women mostly worked in large buildings with large rooms and associated routines of timing and movement. Specific tasks were carried out in specific places and at specific times. The work was concerned with both the needs of clients and the demands of management. The timing of work reflected both these pressures. In most cases the work was very intimate and concerned with the body, whether this be cleaning waste, helping people to get dressed, measuring body size, preparing food and helping people eat, washing and cleaning and whether the clients were present or not. Productivity was the invisible drive in all sectors, whether they were clearly profit driven (the private sector) or had incorporated ideas about efficiency and competition from that sector, as the social care sector had done.[4]

The Social Landscape of Paid Work

Linda McDowell[5] argues that the spaces of paid work are gender coded, reflecting relations of power, control and dominance in the social construction of concepts of work and worker. Such codings also reflect wider social divisions such as age, class, "race" and ability. For example, Rian worked unpaid as a child labourer on her father's farm. She made connections between this experience and her father's account of his own poverty as a child. He was one of eight children in a back to back city house. He had built up a farming empire [*sic*] and escaped the poverty of his childhood. He expected all his own family to work on the farm, even though by the time Rian was born, money was no longer a problem:

> I lived and worked on a farm from a very early age and dad used to pull me out of school to help at hay making time . . . I got the impression looking back on it now . . . that he didn't really value, he did value education, but he didn't, if it was involving his children. He used to want us to work. We were sort of like cheap labour (laughs) and so I never really worked at school. It was just fun. Every school report said, "Has potential. Could do better, but doesn't apply herself" . . . 'cause as soon as I got home I just had to milk cows and stuff and doing homework was just a nonentity.

Rian's experience of work as a child conflicted with a contemporary social concept of childhood in the northern hemisphere as a time for play and learning. For her, the time of childhood was associated with hard physical labour. Ideas about *proper work* and the value of higher education inter-twined in women's accounts of their interactions with significant kin. Rian said of her father:

> But he's a staunch conservative you know . . . and I don't think he thinks that what I'm doing [studies] is going to lead me to a proper job. He sees something that earns money as a proper job and then he says, you know, that the thing about life is to be successful and I sort of throw back at him now all success is not about the amount of money you earn. It's how happy you are and he can see that because he's got older. He's got more reflective. He's got more time now, but he's still such a strong personality.

When Geni told her older sister and uncle that she wanted to go to University, "*my sister, I think she thought it was something everyone else did, so it was good, so she could say, 'Well our Geni's at University.' My uncle, no, because he thought he'd have to pay for it . . . He just thought it was putting off work.*" Her sister had trained to be a nurse and that was considered a more legitimate training and proper occupation for a woman, "*but . . . I suppose that's working at the same time, so you can see the benefits of it immediately can't you?*" Ideas about the value of paid work and higher education shaped women's experiences. The fact that higher education meant current poverty intensified the need for some students to justify their choice. Other people they related to sometimes underestimated the economic need for students to take on paid employment: "*My grant covered my rent and that was it. It didn't cover anything else. People said, 'Why you working and why you doing this [studying]?' You need to eat and live*" (Rachel). *Proper work* also related to whether others chose to pay you for your time. Suraya had worked for several years in community service: *"I'm still doing that now but they're thinking of paying me. I just found that yesterday, for the girls group I'm doing, they're thinking of paying me after three years [both laugh]. I regularly do their play schemes and I go every Saturday to do their girls group . . . and every Monday in the second year when I had free . . . Well just basically doing everything the normal staff do."* She felt her work in the community centre was more highly valued when legitimised with financial payment.

Dominant ideas about femininity and work shaped the work routes which women chose and which were chosen for them. Rian's childhood experience as a farm labourer conflicted with prevalent ideas about femininity: *"When I was working on the farm I struggled with my femininity . . . I wanted to be petite and dress nicely. So when I did go out I was always wearing dresses and showing off my attributes [both laugh] and then I got, I was giving people the wrong messages you know. Oh terrible you know, to be feminine. I wanted to be one of these girls; I was always impressed by these girls who always had a handbag. They had tissues in their handbag."* Beliefs about women shaped her father's ultimate decision to leave his farming business to his son rather than Rian, despite the physical labour he had required from her through her childhood: *"I lived a man's life 'cause I worked like a man. Came in. Was fed by my mother, and did everything that I wanted to do . . . My brother got the businesses eventually. I remember mum saying because he was a man he needed."* When Maura followed an atypical route in terms of paid work and household role, this was conceptualised through gender. She was the main breadwinner with a male partner: *"Any decisions that had to be made in the house, I made them all. It was almost like a role reversal. I was the male."*

The social landscape of paid work was shaped by dominant values and *codings of social division*. Such codings constructed concepts of what was *proper work* for women of different backgrounds and ages. Women's accounts of their experience reflected conflicting mores about childhood, gender, femininity, "race" and paid work. Work was socially valued through financial transactions. Higher education was valued less for some women as it was unpaid. Some patterns of work segregation reflected wider patterns of social division. Separate space was sometimes used for groups identified by colour, origins or ability. Disabled women were consistently discussed as consumers of services, rather than as paid workers. Involvement in paid work directly related to household status and relationships in the household.

The Emotional Landscape of Paid Work

The range of feelings expressed by women in relation to their paid work experiences and their choice to pursue higher education related to both the physical and social landscapes of paid work which they experienced and to past experiences, memories and desires. Suraya hated packing work and associated it with *real work* (see above): *"I didn't like the shift, yeah, cuts and bruises everywhere. It was just horrible . . . It was crates and big things you had to handle. I didn't like it at all . . . but I've got a feeling of what it is like to actually work."* Rian felt far more ambivalent about her experiences of hard physical labour as a child. She found it hard to evaluate her childhood in a positive light because of social concepts of childhood at odds with her experience and also because of her father's fierce control over her as a child. Nevertheless, she felt many aspects of her childhood experience had been positive. Concepts of femininity, motherhood and childhood interplayed in her vivid account of her feelings: *"I suppose work being hard on the farm, and I really appreciated how I used my body physically . . . getting married and having children and not being able to mend fences, knock nails in, put stakes in, climb up ladders . . . I really missed it dreadfully. In fact I would go as far as to say that having to take on a feminine role, having read about it now at college, becoming a wife and mother, did me a lot of harm really."* Rian became quite unwell in her early married life. She felt her choice to pursue more physical activities, including sports, restored her mental and physical well being. This strongly connected in her to experiences of physicality, freedom and constraint as a child.

In social care work, women described feelings of shock, fear and unease regarding the regimes they initially encountered: *"I felt the personal care side of it would be really difficult and that took quite a lot and I can remember the first time I had to wash somebody's bottom you know and I remember that quite clearly thinking, "My God I'm not going to be able to do this." But then gradually it just got easier and I don't think anything about it now"* (Susan). Part of their adaptation involved conforming to the regime: *"You had to change. You were changing to be like them even though you knew that their role was wrong, what they were doing. I mean even if you just sat and had a chat with one of residents . . . you felt that they felt you was not working . . . even though you could see when you were talking to the residents that they were enjoying talking to you . . . Very frustrating. And just the fact that it was like a business rather than caring"* (Irene).

Familiarity with the regime sometimes led to feelings of reliance upon it, and feelings of insecurity in less rigid work settings. When describing the shift from a hospital base to a voluntary sector work setting Rachel reflected, *"I can work on my own . . . I do miss the team . . . and it's a bit bitty. When I was working shifts I knew when I was working and when I was finishing. It's a bit bitty now."* When adapting to the work regime women appeared to reconceptualise the environment in ways more fitting to other life experience, for example, in the family: *"At first it was scary . . . I just kept seeing all these really old people and thinking, 'Life involves, you know, sitting, watching TV. Just waiting to die.'*

But then . . . after a couple of weeks, then you got to know them as people. Then you could see that it wasn't like that, they were just you know all together, because their families couldn't look after them anymore, it was different you know as if you were part of their family" (Geni). But there was a constant tension between the demands of the regime and relationships with clients. This tension was expressed in conflicting emotions, which appeared to be a means of working through the impersonality of the regimes coupled with the high degree of physical intimacy required: "*I enjoy working with the clients very much and you build up a friendship with them and its really nice when you walk through the door . . . They'll turn round and call you by your name, you know . . . and 'How's your family?' . . . You're probably the only person that they've seen all day . . . but from the other side, from the Department, it drives you bonkers. I hate it . . . The more you do the more they put on you to do*" (Susan). At its worst for some women, experiences of paid work were completely negative. Hazel said of her experiences of machining and making women's clothes, "*[It] . . . didn't develop me inwardly to know myself properly.*" This feeling reinforced the personal value she placed on higher education: "*[I] developed a real thirst for it . . . I found my life through education . . . Very self fulfilled with what I was doing.*" However for many women, paid work, even where routine and unrewarding, provided opportunities for friendship and relationships: "*Two summers running I've worked in a bar so, pulling pints, serving food, brilliant. Nice people. Overlooked sea*" (Geni). In this case paid work was conceptualised as a source of pleasure and friendship, features characteristic of leisure.[6] Rashida had not experienced paid work, and felt disadvantaged and a sense of failure, despite the wide range of other responsibilities she held: "*I feel a bit, I should have done that. I should have got a job. I tried for jobs but it just never worked out . . . They were either far away. I couldn't. Transport and that. I tried my best but . . . family responsibilities as well.*"

The whole gamut of emotions informed the emotional landscape of paid work in women's accounts of their experiences. The physicality of work was associated with pain, discomfort but also sensual satisfaction. Intimate relations with others in the paid work setting generated feelings of shock, disgust and tenderness as women made the transition from home-based to work-based concepts of care and love. Some women came to emotional terms with paid work, appreciating the role of management and the structures they had to adhere to. Others remained hostile to the environment. All women were involved in shifting emotions in order to come to terms with the work and to carry out necessary tasks. Below, I outline the key *spatial/temporal representations* associated with women students' experiences of paid work and areas of *emotional dissonance* for particular women.

TABLE 8.1
Space, Time and Paid Work: Areas of Emotional Dissonance

The Dominant Spatial/Temporal Representations of Paid Work	Emotional Dissonance
Through paid work, time has a monetary value.	Negotiating the value of one's time.
Real work makes stressful demands on the body.	Fatigue, pain, and poor health. If work is less physically stressful it may be conceptualised as unlike *real work*.
Clock time dominates work regimes.	Personal and bodily times have lower priority but must be managed.
Paid work usually has an institutional base outside the home.	Work based in people's homes (for example, domiciliary care) may cause particular areas of unease in the management of space and time.
Paid work is gendered and shaped by other social divisions including class, colour, ethnicity and age.	Atypical work routes for women of differing social positions create social unease. Sometimes atypical work for women is termed "men's work."
Paid workers usually have higher status in household decisions.	Level of household control must be negotiated. Sometimes decision making in the household is termed the "male role."
Paid work is impersonal.	This is belied by the brutally intimate nature of much paid work for women.
Paid work is unpleasant.	If work is pleasurable it may be conceptualised as leisure *or pleasure*.
Paid work has a high priority.	Unemployment, whether forced or chosen, may create social unease and personal feelings of failure.

Spatial/Temporal Representations: Housing and Household

The Physical Landscape of Housing and Household

As with paid work, women students' knowledge of domesticity was not confined to the place of their birth or where they lived. Hazel said that her mother was obsessed with the house being clean. When asked why, she said, "*[It] goes back to her childhood [in the Caribbean] . . . No childhood . . . Work, work, work . . . Stolen childhood . . . No time to play. Learning to wash, learning to iron . . . their children just the same. There should be room in life for other things.*" Hazel's mother was ill and her father died during Hazel's course. As the only child at home, Hazel was expected to maintain the rigid patterns of cleaning that her mother had established. Her mother's past was evident in

Hazel's present experiences, and the way the physical environment was managed by them both. The contemporary domestic landscape was very different from the one Hazel's mother grew up in, where the boundaries between domestic and economic production were far more blurred, nevertheless, the values associated with that former landscape influenced the present one. The cleanliness of the home had both actual and symbolic value for many women.

For women students in halls, the dirty environment made it hard to relax and study: *"It is disgusting; it really truly is horrible . . . You couldn't walk barefoot anywhere for fear of treading on cigarette ends, or bits of food . . . It really is grotty . . . I don't think it's got to do with them being in a lower year . . . I just think it's . . . disrespect"* (Geni). Several women spoke of the need for their homes to be clean before they could study and the way cleaning dominated their households: *"I've found that I'm one of these that if I've a piece of work to do and the house want tidied, the housework would have to be done first because . . . with a tidy house I always felt relaxed and I would light scented candles all around the house so maybe it smells nice and make sure the place is Hoover up and tidy and I then feel more relax to sit down and do a piece of work"* (Zandra). Others spoke of the way home-based priorities disturbed their studies. For example, Diane spoke of her mother cleaning around her whilst she studied.

The physical landscape of domesticity was characterised by houses laid out in particular ways in particular housing communities. Within the house, rooms were associated with certain activities. Historically rooms were designed to be mono-functional (see chapter 4). Spatial arrangements and struggles over space inform household negotiations, and, as Sophie Bowlby et al. argue, *"challenges to socially accepted versions of gender often involve the transgression of spatial as well as other boundaries within the home."*[7] When women sought to change the use of physical space in their homes in order to study, their actions were sometimes resisted: *"I was going away from what I call the general sitting room or the dining room . . . and because I was removing myself away from that and finding my own space, that's when the trouble started"* (Mary). Some women did not always feel able to create physical space to study in their own homes. Even where students lived in halls of residence the physical landscape of home and domesticity shaped their perceptions and experience: *"Thing that annoys me is doing everything in the one room . . . studying, sleeping, eating. There's nowhere you can go to get away. Better if a communal area. The only thing you can do is go in a friend's room. But that's just the same . . . go to boyfriend's [room] . . . even if I have to walk there it seems I'm getting away"* (Rose). When describing the way they used the students' union bar, Geni, who lived in halls, said, *"Because we live there, it's just like popping to the shop or something—Not all action really—Molly's gone in her pyjamas and jumper . . . The room in halls is the bedroom, and that [the union bar] is like your living room. There's the telly."* Rose added, *"Going to the Union is like our communal area. It's where we go to get out of our rooms . . . Wouldn't class it as going out out."*

The home is considered a place of private relationships, a *back region*[8] where personal energy is recouped. In some women's experiences however, it was a place where personal energy was drained. The representation of home as a *haven*, a place associated with domesticity and nurture, which are in turn gendered concepts, gives legitimacy to certain activities over others. Women's physical presence in the home led to expectations that their behaviour would be household oriented. For example, Susan said she received lots of support from her male partner in order to study. Nevertheless, he felt that her physical presence in the home meant he could neglect domestic work: *"He used to iron . . . He'd do that for me. But if it was college holidays . . . You've still got all your work to do at home as well as having the children so in a way there's more work . . . and 'cause he thought I was at home and not at college he wouldn't do it . . . So then it used to get really stressful. I used to have even more to do than I would normally and no time at all."* The physical domestic landscape extended beyond the home. The segregation of housing communities by colour, heritage, religion and class gave a distinct character to local communities and institutions. Many domestic obligations were fulfilled outside the home, for example, in negotiations for welfare, health, education and community resources for household members. Local schools had a particular make up and hierarchy of achievement. Students in higher education brought with them messages from school: *"I come from a working class background. I left school at 16 . . . no qualifications . . . very poor academically"* (Hazel). Hazel experienced segregated schooling, having been diagnosed as having learning problems at twelve years. Rashida experienced schooling where there were very few white children: *"'Cause the school that I went to the majority [were] Asian people and I didn't know there's so many people from so many different walks of life you know that you wouldn't normally meet and that's been excellent. What I've found [since college] is nearly 99 percent of the people I've met have been really really nice. You think coming to a white environment there'd be hostility and things like that, but there wasn't."* Spatial representations may reinforce ideologies of nationalism and eugenicism.[9] Rashida identified segregated space as narrowing her experience.

Critical issues in the relation to the physical landscape of home include the following. The relationship between the landscapes that are external and internal to the home is evident in the way the time values associated with the former impinged on the latter. The times of institutions, such as school, welfare agencies and paid work regulated household time, however, household time was multifaceted. The domestic physical landscape was shaped through current and past practices. Strong norms about hygiene, health, sexuality and privacy came through the physical landscape of the home and the physical tasks associated (for example, cleaning). Patterns of movements within the home and community were age, "race" and gender related and resistance was sometimes raised when these norms were contested.

The Social Landscape of Housing and Household

Gendered concepts of childhood and mothering shape the social landscape of housing and household. Hazel's mother's concept of childhood was different from her own (see above). The contemporary representation of childhood as a space and time for play and dependency conflicted with other women's experiences as young carers. Two younger women students who still lived in the parental home took responsibility for less able siblings. Rashida lived with her disabled brother: *"Yeah . . . because we share responsibility for him, my brothers and sisters . . . Can't leave it all to our mum you see. We all help with the bathing, feeding, clothing . . . That takes a lot of time, especially at the weekends. We've got all the housework as well"* (Rashida). Such care responsibilities also involved assuming some responsibility for how their siblings coped and were received in the public sphere. Irene supported a sister with health problems: *"Just little things like, 'Has she brushed her hair?' and things like that. I mean, when I go out with her, I like, sounds awful, but sometimes she just throws anything on that's nearest to her. She's no dress sense. So when I go out with her I like her to look nice. So I sometimes go, 'Oh put this on, put this on.'"* These supportive roles and domestic responsibilities in the parental home related closely to gender and age. Suraya and Rashida discussed each other's housework responsibilities in the following terms. Suraya: *"She's [Rashida] got more [housework] 'cause she's the eldest one really."* Rashida: *"She's [Suraya] more spoilt [laughs] by her parents."* Of her older brother Rashida said, *"The most he can do is lift his plate and put it in the sink. That's what he does. You know he's so spoilt by my parents, especially my mum . . . He's so spoilt even though he's twenty-three years old. He's mummy's boy."* One young woman interviewed took parental responsibility without recognition. Clare had looked after the children of her former woman partner and tried to maintain contact with them. She felt she could not talk about such responsibilities at college, even though this might have helped her negotiate more time to study: *"It's a double thing . . . [a] lesbian relationship and then it's 'somebody else's kids,' not your biological ones."*

The pressure to be *good mothers*[10] affected mother students. Sometimes they had to assert their status as mothers. Sometimes they felt, and were told, that the quality of their mothering was deteriorating because they were students. Mother students received contradictory messages regarding the value of studying from their own mothers. Their mothers created space and time for their daughters to study by sharing domestic tasks and looking after children. At the same time they were critical. Susan said, *"Yes. My mum was a little bit, 'You should be at home with the children.' But it didn't really affect them because when they were off school on holidays I was, and there was only odd times when the school holidays didn't coincide when she would just have them for me."* Beulah's mother was *"really negative about it, so I tried not to rely on her as much as possible."* Lorna got *"a little (support) from my mum, but my mum was thinking that I ought to be at home."* Social disapproval by family members could be punitive. Qasir's mother-in-law, with whom she lived, strongly disapproved of her studies and demonstrated this by *"just not doing the jobs . . . moods, disapproval. Not directly, but telling other people this isn't the age to do it [study] . . . Jobs that could have been done quite easily, you know, putting the washing in the washing machine. Two-minute job. She'd just leave them for me to do. Washing up, when I came home there was a kitchen sink full."* But for younger non-mother students who still lived in the parental home, the support from their own mothers was far less ambivalent. Their right to study was validated through both emotional and practical support: *"I just used to say to my mum, 'I've got these deadlines. Want to get my work done by these dates.' Just asked her to say, 'Oh [Irene] how are you doing? Have you got that done? You said you'd have it done by this date.' And I used to write it on the calendar so. It was more my mum that did that . . . Once you've done it they're all happy"* (Irene). Suraya talked about family support through exam time:

> Exam time, just now when I've had my two exams, I'm really stressed out about them . . . I'd rather do the housework and stuff and they say "No, no just leave that, just go and revise" and stuff. Mum's been absolutely brilliant. She's been praying for me . . . Bringing my food upstairs and . . . I felt guilty at times and I'd sit not doing anything upstairs sometimes when she was thinking I was working. Past three weeks she's been excellent. She has really been supportive. Especially exam time. Always asking me, "How did it go?" and "Don't worry it will be OK if you do this and that and it doesn't matter if you don't pass it."

Some male relatives were supportive of women students. Nevertheless, women still had to renegotiate domestic roles and responsibilities. Such negotiations highlight how landscapes of home shifted and reformed for women. Some men became openly hostile. Male violence was a feature of several women's lives. Kelly, who separated before the course, *"had low confidence, you know low self-esteem with myself. It's like he always used to belittle me all of the time and I realised that and I just thought 'No,' 'cause before I ever met him I was really confident about everything . . . When I met him it just all went downhill."* Hostility and violence could be triggered by attempts to study: *"I was accused of being selfish. I was accused of neglecting my kids. I was accused of being a bad mother. I was accused of not putting anyone before myself apart from me*" (Mary). This experience echoes Rosalind Edwards' research findings[11]: "*The threat to a partner's wishes to have the home as a separate sphere appeared to be demonstrated to the women by the men's reactions to their bringing education home on both the physical [books and papers] and mental levels.*"

The social landscape of housing and household extended into the communities in which students lived. For example, three women of African Caribbean heritage found community-based learning mentors from outside the college to support their learning. Women with children sought space and time from domesticity through community resources, for example, nurseries, other mothers and wider family and kin. The domestic social landscape was shaped through financial transactions, both external and internal to the home. Financial transactions related to time-use are significant in legitimising roles and relationships in paid work.[12] The same is true of home, where the passing of money both values and symbolises that time spent on certain activities is more socially legitimate than time spent on others. In the case of these women students, although time spent in domestic labour was unpaid, financial relationships were central to home life. The state was involved in regulating these relationships, for example, through welfare benefits delivery and the regulation of child support maintenance. The breakdown of financial relationships was sometimes a catalyst in relationship breakdown. Kelly had separated before she became a student: "*Well, we went through a lot, went through a lot. Through the C.S.A. [Child Support Agency] and everything, but they still couldn't trace him which I thought was strange . . . He does maintain them, but not properly. He pays on a weekly basis but not enough to . . . which I think; really he should be more responsible for the children.*" Kelly remembered that her mother had eventually left her own violent husband when he squandered his redundancy pay: "*We always looked nice and clean. She always did what she could for us . . . but we couldn't understand [the violence] . . . My dad got made redundant and he frittered all this money and my mum didn't know anything about it . . . All he bought for us was a bike . . . and he gave my mum just £80. She took him to court then, but almost all his redundancy money had gone by then . . . That still goes through me sometimes. It really, really does*" (Kelly). Critical financial issues also shaped relations with children. Although she felt bad about it, Zandra decided to pay her daughter to do housework in order to study despite this conflicting with her own expectations:

> Sometime she would get fed up . . . and then I would realise, "Oh I'd better make an effort" . . . I'm leaving it too much onto her you know . . . There were times when, although I don't really believe in this, because this wouldn't happen when I was living at home with my parents, but sometime I'd give her some money to

do it . . . So well "Here's five pounds," thinking well at least it will encourage her and she'll think well she's getting something out of it. She can go and spend some money on whatever, yes. I would do that at times. Yes . . . [laughs]. I don't really believe in it but I did it. Because with my own parents there's no way you'd have got money to help out in the house and do . . . do you know what I mean? That wasn't heard of really. (Zandra)

In summary, the social landscape of housing and household was shaped by normative expectations related to age, gender, mother status and so forth. For example, young women who were not mothers appeared to receive more support for their studies from their own mothers than older women who were mothers themselves. Family members were central as a means of legitimising studies. Financial transactions were a part of the process of valuing both household labour and academic studies. Emotion work was intensified for some women students who faced resistance to their studies.

The Emotional Landscape of Housing and Household

A sense of urgency prevailed for one woman with small children, who realised her social and employment opportunities might lessen if she spent the early years of her children's lives at home. Susan decided to study because "*I knew I couldn't have a job because you can't afford the childcare. So I was wanting to use my time effectively and not waste years waiting for him to grow up and then start to go to work.*" Obvious frustrations led Hazel to reconceptualise the value of housework and restructure the work itself: "*I felt it like a burden but my duty. Felt it could be done once a fortnight but she's [mother] used to every week. Wanted it nice for the vicar's weekly visit. 'Vicar's not looking at the floor' . . . Didn't seem dirty to me . . . What I always felt in our house that came first. This is why I used to work early mornings . . . Meeting everybody's needs and leaving my needs behind which was to sit down and do my work.*" For Rashida, housework was an escape from the emotional frustration of studies: "*I tend to do more of the housework than my sisters, but I don't mind doing it because, I enjoy it [laughs]. Sort of a break from all this [studies]. I'd rather do that.*" Rian studied in order to avoid housework: "*You don't do the amount of housework that you used to. It's true. I used to fill my days with housework because that's all I had to do.*" Women expressed guilt, resistance and unease with dominant representations of sexuality and femininity and the association of these with home and nurture: "*Being a wife and mother did me a lot of harm really. 'Cause I ended up having a nervous breakdown and counselling, 'cause I tried desperately to fit into the suburban housewife role and it just drove me nuts and I'd used to been having wide open spaces and then coming here and just having this pocket sized garden*" (Rian). Of young men of South Asian heritage, Qasir said, "*Men have problems relating to anybody on an equal basis. I'm sorry, but the Asian male, you're seen either as a sexual object or a wife, or a mother or sister. Not as a friend or not as someone on an equal basis with you." Clearly these patterns of perception of men's attitudes are common in all communities, not just South Asian.

Concerns about the quality of their parenting affected many mother students. Studying did not accord with traditional concepts of quality mothering: Maggie expressed a *"feeling of guilt because of the amount of time spent away from my son."* Martha spoke of being *"pulled two ways by needing to fulfil my commitments to my study and also spend time with my daughter. The time I do spend with her I am often stressed and in a bad mood with my mind elsewhere . . . I feel that if I gave everything up and stayed at home to be a full time mum I'd go insane and be a crap mother anyway."* Susan said they had *"no quality time . . . they're at school Monday to Friday. I'm at work Saturday and Sunday."*

Two women of African Caribbean heritage expressed strong feelings about the communities they identified with. Similarly to home, it was necessary to resolve conflicting feelings about community as a commitment and a barrier to higher education:

> You find in black communities . . . that a lot of black people in general they won't support you. They won't. I don't know why. I find that things that you don't know . . . People who are willing to help you are there. The support's there but, I don't know. It's just really strange. I've found that in our culture it's just really, really strange people won't support you . . . I don't know if a lot of it's that they don't want you to get to the top and when you've got to top they like to pull you back down again and that's what it's always like, especially in the black community. Especially if you're professional, they'll pull you back down. (Kelly)

However, Zandra recognised as justifiable the community's suspicion of those who wanted to *better themselves*. Would this be at the expense of community solidarity and past friendships?

> I don't like people, who whether they're a doctor, whether they're a lawyer, whether they've got a degree, they think they're better than people. And that just does something to me . . . It gets me really annoyed. So I think, God willing you know, finishing my dissertation off . . . get my degree. I don't want it to change me, I want to be the same person because I don't like it when people go round and think they're better than people because of qualifications and because of their jobs because it really upsets me that.

bell hooks[13] echoes these feelings of ambivalence about higher education as also prevalent in African American communities (see chapter 5): *"Books and ideas were important but not important enough to become barriers between the individual and community participation."* Such feelings carried through into college where some women challenged segregative or exclusionary practices. Kelly talked to a white student who had been ostracised by many, including staff, for racism: *"I thought by doing that with him he altered quite a lot . . . sometimes tutors just back down a lot, where in fact they should get more involved in the actual situation . . . It should have been sorted out there and then in the classroom . . . Because he would have looked at it completely different. No, sometimes you've got to turn it . . . don't think about the black and white issues . . . Just look at a person . . . People just look at the colour instead of the person."*

A whole range of emotions were generated in the landscape of home, and there were key areas of emotional dissonance for some women students. Domesticity may be conceptualised as a haven from paid work and studies but women also expressed the need to escape domesticity. Some women resisted stereotypes that located them in the home and undermined them as students. Although they often felt guilt about mothering, mother students also revalued what it meant to be good mothers (see chapter 10). Ambivalence about responsibilities to home and community created some unease.

Below I outline the key *spatial/temporal representations* associated with housing and household, and areas of *emotional dissonance* for particular women.

TABLE 8.2
Space, Time, Housing and Household: Areas of Emotional Dissonance

The Dominant Spatial/Temporal Representations of Housing and Household	Emotional Dissonance
Domesticity and household time is unpaid.	Financial transactions are central. Negotiating the value of one's time.
Domestic and household time has relatively low social status yet high priority.	The need to re-value and re-negotiate the priority of domestic and household time.
There is a pressure on women to be "good" mothers.	The need to reconceptualise the meaning of mothering and recognise responsibilities held by non-biological mothers.
Within the home, rooms are designed in mono-functional ways.	A need for multifunctional rooms.
It is normative to live in small household units.	Coping with collective living in halls of residence.
Women's commitment to home and community.	Ambivalence about home and community.
The home is a safe haven.	Violence and abuse in the home.

Spatial/Temporal Representations of Leisure

The Physical Landscape of Leisure

Karla Henderson et al. argue that leisure is a commodity as well as an experience, a feeling as well as an activity.[14] The physical landscape of leisure is therefore difficult to convey. Formal leisure tends to be situated in venues outside the home, such as pubs, arts and sports centres although there is also a market for leisure in the home and clearly the home is a site of leisure experiences. The formal places of leisure reflect the divisions of age, class, colour, geographical heritage, gender and community and histories of inclusion/exclusion. Some mainstream leisure activities are structured in similar ways to paid work, that is, by the clock. Susan said she had joined the gym last summer:

> That's my time for me . . . There's still the pressure of time. Still had to get there, get done and get back and I didn't have a shower there. I used to come home and have a bath at home and read and make notes when I was in the bath to save time . . . When the pressure started, I always felt that I shouldn't be there and I used to be there and in the end I had to have a word with one of the instructors because . . . my new programme, it was quite time consuming and I couldn't commit myself . . . He just said split it right down the middle . . . So I just used to go up for an hour . . . Got rid of your stress.

In contrast to the questionnaires, where women cited several activities external to the home (and friendship) as leisure related, in interviews there were different emphases. Betsy Wearing points out that leisure relates to both perception and context and different research instruments generate different responses.[15] Women interviewed were far more likely to mention home as a physical leisure space. For Hazel, her primary leisure was home-based and solitary: "*I read, listen to music. For real escapism, Danielle Steele. Christian books last thing at night.*" Rashida and Suraya, through dialogue express the fluidity of the physical leisure landscape and its connections for them with time at home, time with the family and time alone:

> *Rashida:* I just sit, listen to music.
> *Suraya:* Watch TV.
> *Rashida:* Talk to your family,
> *Suraya:* or just go to my sister's house for the weekend and stay there and then we'll probably go out for the day.
> *Rashida:* Yeah, go out shopping or something or go to the park.
> *Suraya:* Take my niece and nephews to the park when they come over, because they come like every week. I've got a car as well so I can go for a drive around, just clear my head.

The central theme for these young women was that leisure was conceptualised as freely chosen, lacking a rigid timetable and pleasurable. But home was clearly not always conceptualised as a leisure space by women students.

Women also stressed the nurturing aspects of friendship as creating a physical space for leisure.[16] Women living in halls mentioned the need to get out of their rooms, to meet friends and go to the Union bar: *"We were all making friends and getting to know one another. It was a good place to do it [the Union Bar] . . . This year it's become just somewhere we can just get out of our own rooms. Get out from the paper and books all around you, and just go and sit with others, although it's a pressure talking about it at least you're saying how you feel to people that can understand you"* (Geni). Friendships for women students were marked by transition as other students were included in friendship networks and older friendships sometimes faded.

In the context of the overlapping landscapes of home, paid work and higher education, women identified spheres other than formal leisure as leisure sites. Geni and Irene enjoyed bar work, Rashida enjoyed housework and Lorna took pleasure from succeeding at studying: *"I felt wonderful. The very first exam I did within the first year of the degree . . . I passed. I danced all the way down from college. Really happy. Singing at the top of my voice. To the station. And I kept saying 'I've done an exam, in a degree . . . really happy. I was over the moon."* The physical landscape of leisure included important public leisure sites, where activities were available at particular times (for example, the Union Bar in the evening) and to particular tempos (for example, the gym) but the following spaces were also significant sites of leisure for women students. Paid work had the potential to be a site of friendship and companionship. Some women enjoyed studying and attending the college. Home and family were important leisure sites. Being alone was also a source of leisure.

The Social Landscape of Leisure

Women students learned concepts of leisure and its social value from significant others. Rian learned that for women, leisure time may be conceptualised as *selfish time*[17]: *"So my mum was actually running these shops. He'd [father would] be off playing golf . . . and with mum he used to say, 'If you put as much effort into playing bridge ('cause she started playing bridge when older) . . . if you were earning into work (which she did do).' It was almost like she didn't have any leisure time. That she should be working. She worked really hard."* Qasir saw leisure as a form of household and community labour: *"Socialising was a big thing and I just had to say no more. I just couldn't do it. There wasn't enough time in the day. To do the housework, look after the children and study and socialise . . . So I knew that if I continued with the accepted role and played the game I wouldn't do it.*" However, family and community based leisure and ritual could be positively accommodated and enjoyed. When asked how she experienced leisure, Lorna said: "*I didn't. No. Apart from main birthday parties and Christmas, but I was happy for that, because any bit of time I got free I needed.*"

Leisure is often conceptualised as a social entitlement related to employment (see chapter 4). Some women students conceptualised leisure as something to be earned after the prior tasks of paid work, home and higher education. Zandra expressed guilt therefore when she enjoyed herself if other tasks were unfulfilled. She said she sometimes had a bottle of wine, watched television and went shopping with friends when the pressure was on: "*Sometimes I actually did and sometime I shouldn't have done, but I just took myself off into town and I shouldn't have done but I still do [laughs].*" Women had learnt that leisure should not be self indulgent, but a source of health, beauty and fitness. Body image influenced women's accounts of their leisure experiences. Jennifer Hargreaves has written about the enormous pressures on women to be slim, to avoid alcohol and smoking, to remain young in appearance and to be both sexually attractive and sexually active.[18] These representations were often at odds with what truly gave women a feeling of leisure and pleasure: "*I've got sort of more easy with who I am now. But when I'm using my body, it's not, I'm not a sexual person at all you know. That is nothing . . . but for some women, their everyday lives are based around sexuality. I don't know, maybe they're not. Maybe they don't think about it either. Weird isn't it?*" (Rian). Some women appeared to resist representations of beauty, fitness and health but nevertheless felt guilt whilst doing so. When asked how she took her leisure, Geni expressed guilt about her chocolate eating: "*It sounds daft but it's all sort of food based like chocolate or cakes, that kind of thing and then I regret it because I feel how heavy I've got . . . because of spending so much time in the one room. When you come to go to bed you don't feel like it anymore because you've been sat there all day, burning nice oils and having something nice for tea.*" Rose expressed pleasure in *not* getting dressed up to go out: "*Don't have to get dressed up. Fresher's week people get dressed up.*" Leisure has also been conceptualised as a release from the pressure to conform, whether to the demands of home, paid work or higher education and a chance to pursue other identities[19]: "*Last year [first year in hall] . . . a novelty . . . smoking in own room*" (Rose).

Two critical issues for women students in relation to the social landscape of leisure were firstly that leisure was sometimes conceptualised as *selfish time* because the tasks of paid work, home and higher education took priority. Secondly, there were pressures to conform to dominant representations of leisure related to sexuality, femininity and the body.

The Emotional Landscape of Leisure

When asked in interviews how they relaxed from studying, women spoke of activities that were emotionally soothing and comforting to their bodies and minds. Rather than focusing on the things they had "done," they focused on leisure as both a process and a feeling. Women's leisure choices linked to their social routes through space and were shaped by gender, colour, geographical heritage, ability and class.[20] Zandra sought pleasure through an extension of the domestic routine: *"Go shopping with friends as well, lot of window shopping, as a student you don't have a lot of money, but you would go in a cheap shop. See what you can find. Shopping, great stimulation, shopping."* Rian tried to recapture the positive feelings she gained from physical labour on the farm as a child, that is a sense of ease with her own body: *"I started cycling and then I joined a hockey club . . . and then it just went on from there. Being physically active helped me mentally . . . Helped me get some control back in my life and made me feel physically strong and more positive."* When asked how they relaxed, women spoke of time for themselves, their bodily needs and the senses of taste, smell, touch, and hearing. Susan got *"into the bath and shut the door. Sometimes I take some music in and turn the lights out and light a candle or something, if I'm really wound up."* Rian *"studied with music, classical. It used to inspire me . . . I couldn't listen to the radio. I would put music on because the radio would distract me with its jabbering."* These activities were both physically and emotionally soothing. Some women expressed guilt about time spent in such activities. They considered them as not necessarily in their own best interests (health, beauty and fitness) and as taking time from their other obligations (paid work, higher education and household) and as not earned. Rian eventually asserted her right to leisure, whether or not she had earned it: *"I feel that I have a right, as much right as anybody to have my own quiet; my own space."* But for some students the emotional stress, particularly of final year studies, made it impossible to relax: *"Never relax, never switch off, not even in bar . . . Sit and have a laugh. It's always at the back of your mind"* (Maura). The need for emotional support—particularly from other students, was frequently mentioned. Rian mentioned stress. She found it hard to get her mind away from other commitments, and sometimes she felt watched: *"I've been a mixed up cookie . . . 'cause I felt like people were watching me . . . I really did at one point."*

Leisure for women was part of a continuum of life events. Certain artefacts triggered relaxing moods. Both Susan and Zandra, for example, lit scented candles in order to study. Leisure choice was also shaped by memory. Kate refused to go out with friends after the traumatic break up from her husband. She feared her emotions when returning to an empty house: *"I tended to get* the odd night out . . . but then there was the thought of coming back to the empty house and that really upset me and that's why I got to the stage that sometimes I didn't want to go out. I'd rather stop in, because once the door was locked that was fine, but the thought of going out and enjoying myself and then coming back home." Particular environments were associated with both positive and traumatic past life events and women students' choices of space for leisure related to those events. In summary the following key issues arise. Seeking positive emotions, comfort, sensuality, warmth were often at the core of women's leisure choice. This involved intentions to get away from other more oppressive or stressful experiences.[21] Women's routes through space, time and life events informed such choices. Women sought a continuum with pleasant memories from the past.[22] Women also used present spatial representations to situate their leisure choices, within, for example, a web of home-based, paid work and higher education relations. Hence, leisure choice was highly individual and infused with emotion.

Henri Lefebvre[23] has argued, *"The space of leisure tends to, but it is no more than a tendency, a tension, a transgression of 'users' in search of a way forward—to surmount divisions: the division between the social and the mental, the division between the sensory and the intellectual, and also the division between the everyday and the out-of-the-ordinary [festival]."* Below I outline key *spatial/temporal representations* associated with leisure and areas of emotional dissonance for particular women.

Table 8.3
Space, Time and Leisure: Areas of Emotional Dissonance

The Dominant Spatial/Temporal Representations of Leisure	Emotional Dissonance
Leisure as "doing."	Leisure as "feeling."
Leisure as an entitlement.	Concepts of *selfish leisure* and feelings of guilt.
Time for self.	*Selfish time.* Justifying own time spent in leisure.
Clock-led leisure.	Leisure as labour.
Leisure is pleasant.	Fatigue, stress, dieting, alcohol abuse.
Leisure as social.	Leisure as solitary.
Physical fitness, health and beauty.	Leisure as harmful.
Leisure as conformity.	Leisure as resistance.
Formal sites of leisure outside the home.	Home-based activities.
Leisure as distinct from paid work, home and higher education.	Leisure routes in paid work, home and higher education.

Conclusion

In relation to paid work, higher education was represented as a route to better and more autonomous working conditions. It was supposed to increase the monetary value of women's paid work time. However, higher education also temporarily displaced paid work and this was recognised by those whose lives had touched poverty more closely, through their own or kin experience. The experience of higher education was conceptualised as, on the one hand, like paid work, being structured periodically, and involving the production of academic work. On the other hand, for many women, whose lives may have been less personally satisfying, higher education was unlike their experience of paid work, giving them more personal fulfilment.

In relation to housing and household, higher education took women out of the domestic sphere. In this respect for some women it was conceptualised as weakening the household and household relationships. Some mothers in particular internalised these concepts and questioned the quality of their mothering. They also faced resistance to their studying. Younger women usually had their studies legitimated in the household. However, higher education was also conceptualised as benefiting household tasks such as mothering. Such conceptualisations were stressed by women students themselves. In the long term, higher education would also benefit the household economy. The ambivalence about the relationship between housing, household and higher education was also echoed in relation to community. When women became students would they leave their communities behind, would they give back to their communities through their studies or would they form new communities?

Leisure was conceptualised as on the one hand intruding on higher education, taking time away from that more socially valued activity. On the other hand, for younger students, leisure and social activities were essential to their transition to adulthood and higher education is a recognised site for that transition.[24] For older women, higher education was sometimes conceptualised as a site of leisure, being identified as time for self, as is leisure, and providing valued and relaxing space. The concept of *time for self* had negative connotations, as *selfish time*, or positive connotations, as time for personal growth, development and resistance to conformity, hence political time.

A focus on women's emotions gave insight into areas where their personal experiences of home, paid work and leisure were at odds with dominant or normative *spatial/temporal representations*. In chapter 7 I discussed the ways that women's sense of themselves and their social position was produced through the relationship between dominant practices of classification and their specific temporal/spatial position. *Spatial/temporal representations* associated with social divisions were also critical in shaping women's experience of higher education. For example, representations of age appropriate behaviour were highly significant. Mature women students might have felt at a distance from the student role. Particular kin who were not supportive of their right to study because of their age reinforced this. For younger women, higher education often coincided with entry into the labour market and into new adult relationships. The expectations upon them at this particular stage of their lives were intense. Those taking paid work for the first time could be traumatised by events (for example, in care work). The additional expectation that the most appropriate form of housing for young women students was halls of residence seemed to compound the intensity of life changes they were experiencing (as they shared the experiences of forty-plus other students simultaneously in this temporary community).

Women's experiences of higher education were also shaped through class divisions. As discussed in chapter 7, women who identified as working class appeared to have less access to paid work. However, for some of these women, the contrast between their former paid work experiences and their current higher education was so marked that they came to value higher education extremely highly (see, for example, Lorna and Hazel, above). Regardless of class identification, several women felt guilt about their lack of contribution to household finances. Having a financial income was an important negotiating tool in creating space and time to study.

It was notable that all the black women interviewed made reference to the significance of community in their lives, including women of South Asian and African Caribbean heritage, younger and older women and mothers and non-mothers. The history of immigration (either their own or their parents') shaped responses to questions about paid work (Suraya and Rashida) and domestic labour (Hazel). Ambivalent feelings about community came up in Kelly's account of seeking support from the black community, in Zandra's concerns about the potential separation from community that might arise for her if she pursued higher education and in Suraya's references to not being paid for her community work. Although clearly white women's lives were also embedded in particular communities, for the black women interviewed there appeared to be more critical awareness of their relations to particular communities of colour, country of heritage and religion. This awareness carried through to their analysis of racialised relations within education, for example, in Kelly's critique of the college response to racism and in Suraya's discussion of her colour-segregated schooling. The white women, although specifically asked about equality issues in college, made little reference to racialised divisions.

In women's accounts of their experiences, reference to disability issues came through in their discussion of care work, both in the home and in paid work. Disabled people were hence constructed as "other" in the research, even though many of the women students themselves experienced poor health and impairment. Clearly those women had greater demands on their space and time, both in looking after themselves and being forced to justify themselves to others (see chapter 9).

Issues relating to sexuality came through in many women's accounts in the spheres of paid work, home and leisure/community. In relation to paid work, sexuality informed practices of care. In the home, pressures to conform to particular stereotypes were prevalent. In relation to leisure, the focus on health, sexuality and fitness impacted on women's self image and leisure choice.

The dominant *spatial/temporal representations* of home, paid work and leisure shaped women students' lives differently according to their social position, age, heritage and residence. Patterns of meaning associated with different spaces of activity provided their own *imperatives for action*[25] (see chapter 3). The choice to become a student in higher education might be legitimated or barriers might be encountered. For example, some women had no spare time, little physical space, encountered resistance from kin and felt guilty themselves at the time studying took away from their families. However, because of their complex and different personal and social histories, the types of barriers particular women faced were not predictable. Neither were the ways in which such barriers were transcended: *"Space is at once result and cause, product and producer; it is also a stake, the locus of projects and actions deployed as part of specific strategies, hence also the object of wagers on the future."*[26] In chapter 9 I focus on women's experiences of space and time in the college and in chapter 10 I explore the ways that women use space and time to benefit their academic studies.

Notes

1. Arlie Hochschild, "The Sociology of Emotion as a Way of Seeing," in *Emotions in Social Life: Critical Themes and Contemporary Issues*, edited by Gillian Bendelow and Simon J.Williams (London: Routledge, 1998), 11.

2. John Urry, "Sociology of Time and Space," in *The Blackwell Companion to Social Theory*, edited by Bryan S. Turner (Oxford: Blackwell, 1996), 163, citing Giddens.

3. Luke Clements, *Community Care and the Law* (London: Legal Action Group, 1997).

4. Clements, *Community Care*.

5. Linda McDowell, *Capital Culture: Gender at Work in the City* (Oxford: Blackwell, 1997).

6. Eileen Green and Sandra Hebron, "Leisure and Male Partners," in *Relative Freedom: Women and Leisure*, edited by Erica Wimbush and Margaret Talbot (Milton Keynes: Open University Press, 1988), 84.

7. Sophie Bowlby, Susan Gregory and Linda McKie, "Doing Home: Patriarchy, Caring and Space," *Women's Studies International Forum* 20, no. 3 (1997): 343–50, 346.

8. Urry, "Time, Space," citing Giddens.

9. Paul Connolly, "Racism and Post Modernism: Towards a Theory of Practice," in *Sociology after Postmodernism*, edited by David Owen (London: Sage, 1997); Paul Gilroy, *Small Acts: Thoughts on the Politics of Black Culture* (London, Serpent's Tail, 1993).

10. Elizabeth de Silva, *Good Enough Mothering? Feminist Perspectives on Lone Motherhood* (London: Routledge, 1996).

11. Rosalind Edwards, *Mature Women Students: Separating or Connecting Family and Education* (London: Taylor and Francis, 1993), 111.

12. Karen Davies, *Women and Time: The Weaving of the Strands of Everyday Life* (Aldershot: Avebury, 1990), 26, citing Thompson.

13. bell hooks, *Talking Back: Thinking Feminist—Thinking Black* (London: Sheba, 1989), 89.

14. Karla A. Henderson, M. Deborah Bialeschki, Susan M. Shaw and Valerie J. Freyysinger, *Both Gains and Gaps: Feminist Perspectives on Women's Leisure* (State College, PA: Venture Publishing, 1996).

15. Betsy Wearing, *Leisure and Feminist Theory* (London: Sage, 1998), xii, 25–37.

16. Eileen Green, "Women Doing Friendship: An Analysis of Women's Leisure as a Site of Identity Construction, Empowerment and Resistance," *Leisure Studies* 17 (1998): 171–85.

17. Eileen Green, Sandra Hebron and Diana Woodward, "Women, Leisure and Social Control," in *Women, Violence and Social Control*, edited by Jalna Hanmer and Mary Maynard (London: Macmillan, 1987); Rosemary Deem, "Feminism and Leisure Studies," in *Sociology of Leisure: A Reader*, edited by Charles Critcher, Pete Bramham and Alan Tomlinson (London: E. and F. N. Spon, 1995).

18. Jennifer Hargreaves, *Sporting Females: Critical Issues in the History and Sociology of Women's Sports* (London: Routledge, 1994).

19. Wearing, *Leisure, Feminist*.

20. Davina Cooper, "Regard between Strangers: Diversity, Equality and the Reconstruction of Public Space," *Critical Social Policy* 18, no. 4 (November 1998), 465–92.

21. Susan Shaw, "Conceptualising Resistance: Women's Leisure as Political Practice," in *Journal of Leisure Research* 33, no. 2 (2001): 186–201.

22. Sheila Scraton and Beccy Watson, "Gendered Cities: Women and Public Leisure in the 'Post Modern City,'" *Leisure Studies* 17, no. 2 (April, 1998): 123–37.

23. Henri Lefebvre, *The Production of Space* (Oxford: Blackwell, 1991), 385.

24. Claire Wallace, "Between the State and the Family: Young People in Transition," *Youth and Policy* 25 (1988): 25–36.

25. Jurgen Habermas, *The Theory of Communicative Action. Volume Two—Lifeworld and System: A Critique of Functionalist Reason* (Cambridge: Polity Press, 1981/1987).

26. Lefebvre, *Production, Space*, 142–43.

9

The Framework and Guidelines of Higher Education

Introduction

"What though does it mean to talk of 'safe' space for women? Safe might mean 'safe to': safe to explore women's lives and experiences; safe to explore constructs of knowledge and safe to explore structures of power. But safe can also mean 'safe from': safe from sexual harassment and from violence; safe from constructions of gender that so often limit our life opportunities; safe from being silenced or having to act like men in the patriarchal mode of the academy."[1]

IN THIS CHAPTER I draw principally on student interviews[2] in order to discuss the *spatial/temporal practices and representations* of higher education which women students experienced. I discuss *snapshots* and *landscapes* of experience, following the format adopted in chapters 7 and 8, but in this case, rather than focusing on women's experience in the spheres of home, paid work and leisure/community, I directly focus on the institution of higher education. The complexity of women's experience means that a degree of overlap is unavoidable. The chapter is in three parts considering: *the spatial/temporal practices* of higher education: *snapshots* of women's experience; *the spatial/ temporal representations* of higher education: *landscapes* of women's experience; conclusions: experiencing space and time in higher education.

Chapter 9

Spatial/Temporal Practices of Higher Education: Snapshots of Women's Experience

In this section I address the research question raised in chapter 5: *What are the dominant spatial/temporal practices of higher education shaping women's experiences?* The section is structured around critical themes in relation to the timing, organisation and culture of contemporary higher education. Of the students interviewed, Qasir was the most forthcoming in her critique of contemporary higher educational practice. She was involved in a close-knit circle of women student friends who gave each other support. Qasir said she gained a lot from her studies and was positive about the curricula and aspects of tutor support. However, she raised critical issues concerning her experience of negotiating space and time to study in the college:

> Once you got hold of somebody you were OK, but getting hold of somebody was difficult at times, especially the interim periods, e.g. the three week break at the end of the January . . . If there were things going on, or if you hadn't made the submission date, getting hold of someone became a real problem . . . A lot of people in the group used to talk about why can't they have one contact person . . . and then that person to try and contact the appropriate tutor (even a secretary). I understand that you can't give everybody's telephone number. But if you had a contact person, ring them up and say there's this particular student looking for you can you make sure you get in touch with them . . . So again a lot of people felt that although the staff were quite supportive, there was a gender issue . . . They received a lot more support from women members of staff than they did from the men.
>
> I think Christmas was worse, because it's a nasty time of the year anyway and I mean I'm completely non-religious. I don't believe in religion full stop and all the associated rites and rituals, but children get caught up with it and it's the time of year when everybody has got some form of holiday . . . It's a time when people seem to visit for some reason more than they do normally. So again that was very difficult . . . The kids demand their ideal Christmas which is fed to them through their school and the visitors. A lot of the assignments were due in very early in January and the other thing that I used to complain about every semester to the library, the timing of the library. Because the library actually keeps term time and the college keeps semester time so the library closes fairly early on. So you haven't got any time. If you don't get your books out, you haven't got any time. You've got no recourse over the holidays and that's what Easter and Christmas [are like].

The following themes emerge from analysing Qasir's comments:

- The rigidity of temporal rationales informing higher education practices, for example assessment deadlines.

- The impact of semesterisation and the dysfunction arising from the imposition of a semester calendar over a term-based calendar.
- The collective nature of higher education and increased student numbers leading to changing demands on lecturing staff and difficulties in access.
- The gender and other social divisions shaping the experience and structures of higher education.
- A higher education culture which emphasises individuality, enterprise and production.

This *snapshot* of higher education highlights the manner in which the higher education *system's imperatives*[3] have impacted women students. Below I discuss each of these issues in turn, showing where other women students raised similar points.

The Rigidity of Temporal Rationales Informing Educational Practice

In chapter 5 I discussed timing as a central form of regulation in higher education.[4] This timing follows the process of teaching, assessment, moderation and evaluation within academic institutions. In the case of these women students there had been an intensification of the assessment. They were expected to produce a greater number of assignments after semesterisation. This created a log jam effect at the end of each semester. Rashida and Suraya referred to the limited time between exams, which meant that they had less time for revision. Rose and Susan felt that the exams began too soon after essay assessment deadlines. Hazel said, *"It's a bit ridiculous. Why can't they sort it out? Mountains of essays to be in all at the same time. Why can't they sort it out?"* Students needed to negotiate tutorial time: *"Worried you were encroaching on somebody's time . . . It isn't always an appropriate time and that makes you feel bad"* (Rian). Negotiating extra time involved coming up against tight regulations on submission of work. Late work required mitigating circumstances. Mitigating circumstances required sick notes. To get extra time for assessment, students needed medical evidence: *"Wasting my doctor's time when sick people should have had appointment . . . One piece of work I didn't have time to finish . . . A hard piece of essay . . . Everybody was finding it difficult, and I was reading, reading, reading, making notes, making notes and really putting the effort in but I just couldn't finish it for some reason"* (Zandra). Recognised impairment or illness was a route to extra time. Kate had epilepsy but was not aware of her rights until too late: *"One thing . . . that I didn't know anything about the fact that we should be registered as disabled . . . regarding the epileptic fits. I should have been told about the [learning support] centre at [the college] and there would have been mitigating circumstances all through the three years that we were here . . . I found out that the drugs that I was taking for epileptic fits do tend to . . . not numb your brain, but [slow you down].*"

Semesterisation

The move to semesterisation has the potential to allow faster degree routes for students. Three semesters can be fitted into one academic year and thereby facilitate two-year degree programmes. This may be a long-term policy aim in the U.K. In the case of these women students, the semester timetable interrupted the traditional academic calendar in confusing ways that were sometimes detrimental to their higher education experience. Although not recognising that their problems arose from semesterisation, students complained that there were "*too many breaks*" during the course (Suraya and Rashida). Holidays were often followed by semester breaks, which were busy times for staff involved in marking essays and preparing teaching for the following semester, but which were unstructured times for students. Rian felt that her third year of studies was more satisfying. This was because all modules in that final year carried through two semesters.

The Collective Nature of Higher Education and Increased Student Numbers

Of necessity higher education is involved with groups of students rather than being tailored to individual needs. This is cost effective. However, increased recruitment of students without corresponding increases in staffing had intensified the process for these students. *Staff/student teaching ratios* had increased. Strict demarcation of tutor roles, as year, course and subject tutors for instance, was not fully understood by students. Also the continual development of new educational and management practices led to inconsistency between staff. This impacted on students in several ways. Dealing with a variety of tutors, Geni felt "*passed around . . . not knowing where to go.*" Rian felt that sometimes approaching the *wrong* tutor resulted in a negative response. Lorna confirmed this. When she approached the *wrong* tutors, she said, "There'd be a couple of teachers that would look at me stupid. 'What are you doing here? You're wasting my time!'" Mary said that she felt mature students were particularly pressured to take a more leading role in class discussions because of the large numbers of students.

Social Divisions

Social position was critical in determining the ways in which *the spatial/temporal practices* of higher education impacted on different women students. Social position, in turn, may have been partially constructed and certainly reinforced through such practices. For example, the rigid temporal rationales of higher education were particularly difficult for women with multiple commitments, for example mother students (Susan), or whose time was squeezed for reasons of their own health or abilities (Kate). Semesterisation and the confusion arising from the imposition of two different calendars, one of which was based on Christian holidays, impacted on students who were involved in faiths other than Christianity. Muslim women found themselves revising for exams during Ramadan for example (see chapter 10). Large student numbers without respective increases in staffing created pressures on physical space and made access more difficult. For example, large numbers in classes disabled women who had hearing impairments and dyslexia (for example, Rachel) and who found it difficult to concentrate in order to take notes.

Individuality, Enterprise and Production

In chapter 5 I referred to the impact of enterprise culture within higher education.[5] As with all paid work situations, staff are torn between the demands of management for greater productivity and efficiency, and the needs of consumers, in this case of students, for a good learning experience. Restructuring of practices in higher education institutions is ongoing. The concept of the *enterprising student*[6] may have some subtle benefits for students when used in empowering and student centred ways. It may be a means of valuing the skills and knowledge of individual students and facilitating self directed learning. But at the end of the day the *enterprising college* takes precedence over the *enterprising student*.

This particular cohort of students was badly affected by the merging of college campuses and the closure of their main site of study. The management rationale for the move was that *staff/student space ratios* on this particular campus were too big and that the merging of campuses facilitated more efficient teaching in the college as a whole, as well as cost reduction. Although the final closure of the campus took place in the summer break, in reality, preparations for the closure had effects all year. There were staff reductions and relocations. Students complained that books were removed from the library and sold when they were still revising for exams: "*Towards end . . . books moved, sold*" (Rose). "*Also impact of move, as books were moved around. Panic!*" (Geni). They felt demoralised and unsettled by the process. Although the issue of campus closure was unique to this particular cohort of students, virtually all higher education institutions are affected on an ongoing basis by major restructuring in pursuit of so-called efficiency and cost reductions.

Summary

Women stressed problems arising from several (often interrelated) factors. There was pressure to produce assessments to rigid deadlines which were rarely staggered. They faced difficulty negotiating extensions and deferring work unless there were medical grounds. The overlapping timetables created lengthy and apparently irrational gaps in the programme of studies. They faced problems with access to appropriate tutors and confusion over tutor roles. Large student numbers had increased *staff/student teaching ratios*. College-wide restructuring was an attempt to reduce *staff/student space ratios*.

Spatial/Temporal Representations of Higher Education: Landscapes of Women Students' Experience

In this section I address the further research question from chapter 5, "*What are the dominant spatial/temporal representations of higher education shaping women's experiences?*" I consider the physical, social and emotional *landscapes* which women students experienced at the college.

The Physical Landscape of Higher Education

Although all the spaces in the college were designed to function in specific ways, women used them in a variety of different ways, transforming them into multifunctional spaces. This was an important part of the higher education process. The college was in a small town on the edges of a large industrial conurbation. There was a large central building containing a security office, reception area, boardrooms, staff rooms, a computing facility, offices, a library complex, a hall and some teaching rooms. The building had large rooms with high ceilings and although in need of repair and renovation it was a spacious environment for teaching. There was some student accommodation on upper floors. There were three adjacent buildings. Two were the focus of different areas of study and one housed the students' union. The students' union complex included a bar, meeting rooms and office space. The campus was wooded, with flowerbeds, grass and a lake where large greedy ducks congregated. The college overlooked moors and occasionally sheep wandered onto campus. The site was easily cut off in bad weather as it was at the top of a steep hill. The central building was erected in the last century and designed with certain normative expectations in mind, in particular that only able-bodied people would need to access it. Physical access was difficult for staff and students alike. Lift access was limited to parts of the building: "*In the second year I had problems with my knees. Was on crutches for four months on and off. Living at the top of hill. Getting to lectures and placement to do as well. Even now I'm just absolutely disgusted with the lack of support"* (Clare).

The college contained space for living. Many women lived in halls of residence close to and on the college campus. Halls consisted of single rooms in wings and on upper floors of the main building. There were also converted cottages in the grounds. Students tended to associate with other students they lived close to in the early months of the course. There were kitchen, bathroom, and washroom facilities in each area of residence. Halls of residence are often represented as a positive, safe and supported environment for younger women who have just left home and many of the younger first years lived there. The experience of support from peers in this residential setting was vital for many young women: *"In the first year it seemed like a great big new experience... You had your friends round you, no matter day or night there'd always be someone, you know you could go in to or go and sit with... We always did things together ... all cook for each other and go out together, watch videos together. Everyone used to joke because we'd all be sat round with cups of tea and one biscuit tin"* (Geni). However, halls of residence also replicated many of the features of institutional life well critiqued in social care literature.[7] Women in halls of residence had conflicting feelings, especially after three years there. On the one hand they felt a lack of privacy, sometimes to the extent of feeling under surveillance. On the other they felt anonymous and isolated: *"Like an open prison. Got your little cell. Can 'to and fro' and go downtown if you want"* (Rose). Women in halls of residence faced difficulties with studies unique to the physical setting they lived in. Rose described the intensity of both living and working in a small physical space filled with noise: *"It's really annoyed me—noise and that... dissertations, essays, revision. I was working then from morning right through and I was stressed. Had to tell group of students, top room... Play station, 'Do you mind if I shut the door?' Still noisy, but at a level... try to leave it for half an hour and after that... Music from upstairs... downstairs... light shakes... At night, once I'd done my work... all my books, files around... Couldn't get it out of my head. Felt like you couldn't escape work"* (Rose).

The delivery of the curricula took place in a variety of spaces in the college. Efficient delivery to large numbers required the use of large lecture theatres for some of the teaching. Generally, women preferred being taught in smaller rooms with fewer other students. They said they found it easier to concentrate. The anonymity in the large hall meant students were less likely to feel confident or to be called on to participate, and they often *switched off*. Susan felt *"very self conscious you know when you saw the faces."* Irene *"didn't like it. Daren't speak in there."* Rian *"felt talked at."* Kelly and Rose could not concentrate: *"They'd be talking about a certain thing and then part way through they'd change it and I've lost it and put my pen down"* (Kelly); *"Don't learn well from big lecture hall. Couldn't put my hand up and ask. More likely to drift"* (Rose). Maura said she used the time in large lectures to work out her household finances: *"It's an escape for me I think . . . You sit in a room and you can do whatever . . . I just switch off."*

The library was sometimes used for personal study and some students enjoyed the environment, such as Geni who said she *"loved books."* Others felt more intimidated or distracted. Rose *"only sat down [to read] reference books."* Maura was distracted: *"I feel distracted. People come in, 'Oh just let me tell you this' . . . I just used to think that I didn't fit . . . I just saw the library as a work place."* The silence rules in the library made it an opportune place for sharing secrets. For example, as well as studying there, Suraya and Rashida used the library *"to share personal issues."* Academic texts could be intimidating. Rachel, who worked as a nursing assistant, felt that her years of experience were not valued in her workplace. Developments in Nursing had brought in more staff on higher grades and with degree education: *"Nurses that'd write the ward round sheets . . . some of them would . . . 'know better than you' because it 'comes out of a book.' No client common sense at all."* For Rachel, academic texts were perceived as not drawing on real life experience. Suraya was shocked that the books in the library had no pictures. Her only previous experience of libraries was at school.

For some students accessing and learning the ever-developing technology was an intimidating process. Two spoke of very negative experiences and lack of support in the computer room. Rian *"[didn't] feel welcome. Made you feel like you were interrupting."* Rose *"came out nearly in tears. Stood there for ten minutes while two assistants talked to each other . . . then said 'busy' and walked off."* Maura felt distracted there and could not concentrate. She said other students constantly interrupted her: *"They all want to know the gory details. 'Yes . . . we've split up . . . No, he's not committed suicide.' That's why I bought my own computer."* However it was evident that many students made gains once they had accessed and learnt how to use the equipment, particularly those supported through the learning support unit, for example, Kelly.

Clearly many parts of the college were used for socialising (see below). Students who lived in halls of residence saw the students' union as *their* place to relax and escape from their rooms. However, some women found the union an uncomfortable building: *"I did try it out. Felt uncomfortable. Different culture again. It would be good if there were more black students"* (Kelly). Suraya and Rashida responding to each other, made the following comments: *"Went once or twice. Didn't like atmosphere. Dark and gloomy. Empty. Nothing to do. Bar."* But when asked if they would have used a separate alcohol-free space, they said, *"Different room for use would be like segregation. We're not used to an* environment like that. We joined the ethnic minority group. Once I went to that meeting but that was ridiculous . . . That was like segregation in a sense . . . we found it inappropriate . . . That's the whole point of us being here, integrating." The group they referred to had emerged out of the student process of challenging racialised practices in the college. For these two young women however, its very existence represented that segregation was still prevalent. They saw its existence as reinforcing rather than challenging racism.

Some spaces in the college I have termed *in-between*, for example, the canteen was a large space with tables and chairs seating groups of eight to ten people. In addition to meals, it was used as a space to discuss teaching and learning as well as a space for friendship and companionship. Lecturers sometimes sent students there to work in small groups when there was not enough room elsewhere. Rachel *"used the canteen for chuntering."* Maura found the canteen *"very daunting . . . massive . . . used for socialising and [preparing academic] presentations."* Both Kelly and Rose said that although they never ate in the canteen (because they did not like the food) they enjoyed sitting there in the company of other students. The smoke room was a fairly small space with one window and very poor air ventilation. Clearly smokers gathered there, but the occasional non-smoker, such as Kelly, also used it. The smoke room was used for both socialising and academic work. Many students were observed sitting writing out essays there rather than in the library. There were designated student common rooms for two specific student cohorts. Sometimes students outside these cohorts considered them *cliquey* spaces. Maura used the student common room to thrash out ideas: *"Common room . . . I love it. A social space. You can have a coffee. Way of catching up. If there's something to talk about from a lecture . . . People say things they'd never say in the lectures . . . I used to love sitting, arguing. Way of sorting it out in my head as well."*

A critical aspect of the physical landscape of higher education for women students was the accessibility of space for their studies. Higher education was a process that took place in a variety of spaces. Some of these spaces were not intended for that purpose and therefore no formal learning support could be found there. Some spaces that were specifically designed for teaching and learning were considered inadequate. Spaces for living and socialising had different potential and limitations in terms of academic studies.

The Social Landscape of Higher Education

Women brought expectations from their communities to the higher education institution. Kate had internalised the message, reiterated by her son, that she was too old to study. When she joined the students' union, this feeling was uppermost in her mind: *"I was actually joining the students' union, and there's* these two lads . . . in leather gear . . . earrings and everything and I'm thinking, 'Oh my god.' So it was like a new world to me and there was this young girl there . . . and she said something to me and I said, 'I know you, do you know me? . . . Do they call your mum [Barbara Rollings]' and she says, 'Yes.' I says, 'Oh my God . . . what's up . . . bloody hell. I went to school with your mother.' I says, 'What the hell am I doing stood here?'" In fact Kate said she found great support from young students who legitimated her right to be there: "Well, these two lads turned round laughing. They said, 'What's up with you? You'll be alright.' They were the best friends I've ever had. Fantastic."

Some women felt more confidence in their right to study but faced challenges from other students.[8] Clare faced harassment as a lesbian, and said, "Before college there was very little harassment for who I was . . . for my identity. Then, it was the whole thing about expecting to find a community college, equal opportunities, you've got every type of oppression. But I think it's harder to deal with that in college because it's underhanded stuff. It's not blatant in your face stuff, which you can deal with." She found her right to occupy particular spaces in the college was challenged: "There's a big thing especially around gay students . . . you kept separate. The rugby lads were fine. It was the football group. There's this bizarre thing around . . . It was guaranteed if ever there was a group of male or female people who they knew to be gay there'd always be a comment or an argument . . . It was really bizarre." Sue Jackson[9] has also written about the difficulties women have in finding safe space to study: "The students in my study have described harassment, being silenced in the classroom and finding their experiences devalued or ignored."

The curricula both reproduced and challenged social divisions and women students responded to the curricula differently. The curricula had emerged from a set of practices and academic disciplines that gave visibility to some issues over others. The emphasis on community, gender, health and social care (the chosen academic and career routes of many women at the college) meant many students responded to it positively and most comments on curricula in interviews were positive. But there were key exceptions. Although most women students welcomed the emphasis on social divisions in the curriculum, critical weaknesses were identified. Diane mentioned a lack of consideration of issues related to Northern Ireland and Zandra, a lack of full consideration of black perspectives across all parts of the curriculum. Issues of masculinity were not sufficiently addressed (Diane) leaving some men on the course feeling labelled as "*caring and namby-pamby.*" Lorna, Mary and Clare said homophobia was not adequately challenged in the classroom. Clare said that complex issues related to sexuality and extreme religious beliefs were inadequately handled: "*Things we were being taught in lectures compared to the issues we were dealing with were worlds apart. We were discussing things around sexuality and you've got some people who've just never heard the word bi-sexual . . . think that means she's gone with a man and woman at the same time . . . At the other end you're facing someone really threatening you. How do you relate the two?"* Clare demonstrated the way higher educational practices tended to reinforce normative assumptions even when critical issues of social division were being addressed. Some students felt that their learning experience was weakened as a result. The fact that they were dyslexic had to be repeatedly mentioned by students to staff, even though they had been fully assessed and policies put in motion (Diane).

The dominant representation of higher education was as a sphere for gaining knowledge. Some women therefore felt a lack of confidence in what they already knew: *"Sometimes you feel a bit stupid and you daren't say anything because you did feel that all the tutors here were really clever . . . But in the end I just gave way to that and thought, 'I don't care whether it sounds stupid or whether it sounds naïve, I'm just going to ask because if I don't ask . . . Just ask, ask, ask now.' For a lot of people they have said they've really been intimidated"* (Diane). The higher education culture, which positioned staff and students hierarchically in relation to knowledge production, reinforced the belief of some students that their knowledge and understanding had less validity.[10] However, other aspects of college culture sought to undermine these tendencies. The teaching was recognised to be of high quality, both encouraging participation and validating students' knowledge.[11] Methods that valued students' prior knowledge and experience were also valued by women students, in particular those that, as Kate McKenna argues, did not impose a too explicit conceptual framework on women's experiences.[12]

In summary, women students' experience of the social landscape of higher education came through their interactions with and particular social expectations emanating from: other students, both individuals and groups; academic staff; and the curricula and learning environment. Through these interactions women encountered social barriers and social support to their individual learning.

The Emotional Landscape of Higher Education

Two general themes are discussed in this section. Firstly, in line with Joan Parr's findings, the higher education landscape was perceived through critical personal life events,[13] which were associated with strong emotion and still reverberated in women's lives. Motivation for entering higher education was often bound up with these events. These were experiences related to relationships, money, paid work, housing and schooling. These memories recurred in women's accounts of their experience of higher education. Diane had vivid memories of trying to support her mother as a child after her father had beaten her: "*I remember bundling my mother into a taxi when she was so badly beaten, taking her down and saying she fell on a door, she fell on a doorknob. I had to explain to the GP because my mother wouldn't admit to being beaten.*" Rian relived memories of depression and anxiety when she was at home with small children: "*I felt so restricted. I felt like I was tiptoeing through life. Instead of striding out in life . . . I just wonder how many other women will go through the whole of their life like that.*" Women's feelings about their studies were connected to such experiences and emotions. The decision to study emerges from past experience,[14] for example, it may involve challenging the low expectations of other people; rejecting a certain lifestyle or seeking autonomy for self because of relationship breakdown. The emotional landscape of higher education was shaped through these memories and desires.

Secondly, the emotional landscape of higher education was shaped by women's ongoing struggle to find space and time to study. This aspect of the emotional landscape of higher education was well conveyed by Lorna, who spoke of trying to enlist her male partner's support to write an essay that was due in the next day. There were clear links between her attempts to negotiate space and time at college and at home:

> I'm saying, "It's this point, that point, can you just help me? Does this question mean blah blah blah?" He said, "Yes, I'll help you" and basically was talking just about one thing but forever and ever. I'm saying "I haven't got time. No. Can we talk about this?" But he wanted to be in control you see. Just talking about anything and everything. And I said "I've got to get it done tonight." Then he said something else which was totally wrong and I realised then he was talking about anything just to calm me down so he could go to bed. And I said "No, it doesn't mean that." And because I criticised him he ranted and raved and he says "Do you know you shouldn't even be on a degree course . . . You don't even understand the flaming question . . . [Your daughter] can read better than you." He went to bed and this was ten o'clock and I thought, "I don't understand the question. I don't know what I'm doing and I've been told and I don't understand it still" . . . and then I thought, "No, I've done it so far." I worked all night till six o'clock in the morning. I finished it from ten o'clock at night, crying for most of that because I was still upset. Typed it up.

Lorna's relationship with her male partner had been full of conflict. His support of her studies gave him opportunities to assert his authority in the relationship as both considered him to be *cleverer* than she. Lorna's growing recognition of the inadequacies of his support and her criticism of this, led him to withdraw academic support and to insult her intellect. For Lorna, hostility at home reinforced her emotional engagement with college. She expressed this through referral to *home* as a feeling associated with college and not where she lived: *"I loved the college atmosphere . . . It felt like home. Well it was my home. I didn't want to come home, apart from the children . . . For a week while I was doing my exams in the third year I stayed in halls. Although I missed my children I certainly didn't miss [male partner] at all. It was just a big weight off my shoulders and I loved it at college."* She said that her confidence was transformed through succeeding in higher education.

Rian's partner was generally far more supportive of her studies. Nevertheless, tensions arose when she raised feminist issues or appeared to be enjoying the course too much:

> I was very enthusiastic about college and I loved what I was learning . . . and I would talk about it with [male partner] a lot and we would always end up in arguments . . . We'd have to agree to disagree. But I always wanted him to sort of like, I really wanted to give him this information, "Look at this, somebody's written about this," opening doors, and he just wasn't interested and still now I never know why. Whether it was . . . because he didn't want to learn from me. Whether it was a male thing, you know? Or, and he's thrown quite a bit of what I've learnt back in my face. I've made him aware of a lot of issues that he would rather have been in ignorance about. And it's caused quite a lot of problems, and hasn't always been easy.

Emotion shaped motivation to study from the outset. Memories of emotion shaped current experiences. In addition, emotions acted as catalysts of change in women's relationships with others. Attempts to study and to include others in the process often triggered strong emotion and unsettled balances of power. Initial fear and insecurity might be replaced by enthusiasm and exhilaration as barriers were overcome by women students. Initial encouragement from partners might fade away as the domestic pressures built on them over the years of the course. The emotional labour involved in such situations was intense. Critical aspects of the emotional landscape of higher education were: enduring emotions related to past critical life events through which present studies were mediated; gendered negotiations and conflict around space and time to study; conflicting pressures on women from home, paid work, leisure/community and higher education (see chapter 8). In table 9.1 I outline the dominant *spatial/temporal representations* of higher education as experienced by women students and areas of emotional dissonance for particular women.

Conclusion

The concept of *spatial/temporal practices* has given insight into the impact of the contemporary routines of higher education, such as assessment procedures

TABLE 9.1
Space, Time and Higher Education: Areas of Emotional Dissonance

Dominant Spatial/Temporal Representations of Higher Education	Emotional Dissonance
Higher education provides accessible space for learning.	Boundaries and barriers encountered.
The higher education institution is a site of learning that contains particular spaces for living and socialising.	Such space is used in multifunctional ways and the practices of living, learning and socialising overlap.
Certain formal lecturing styles are still prevalent, in particular for large groups of students.	A preference for less formal lectures, more interaction, smaller groups, and processes of learning and knowledge construction linked to life events and emotions.
Hierarchies of status in the institution in relation to knowledge production.	Students' self esteem.
The ideal of residential living, particularly for first year students.	The impact of collective living.
Higher education is a recognised site for the transition to adulthood.	Mature students may feel out of place.
Higher education is a relatively safe, tolerant and inclusive space.	Harassment/exclusion of particular groups.

and timetables. The concept of *spatial/temporal representations* has given insight into the ways that physical and social space in the college was laid out and the barriers and support which women encountered. Through this analysis it is possible to see the way that women students' experience was shaped by particular *system imperatives* emanating from: national policy, in particular the emphasis on productivity, enterprise, and efficiency, increased student numbers and cuts in finance; the design and function of space in the institution; social expectations conveyed through the curricula and interactions with staff and students.

Some of the pressures on students were clearly inevitable as they took on the additional work arising from their choice to pursue degree courses. However, some pressures arose from a combination of factors related to higher education practice. The way that higher education was laid out (physical and social space) created areas of *emotional dissonance* that could interrupt the learning process. In addition, relatively recent policy changes at a national level had begun to transform the teaching/learning process. In particular, a *business ethos* was evident in women's accounts of their college experience. Hugo Radice[15] argues that *"The Business University"* has now taken over every campus in the U.K.: *"For the university has indeed been 'reduced' by financial stringencies, authoritarian management, political intervention, creeping privati-*

sation, and, overall, the almost complete commodification of teaching and research." He points to the lack of long-term planning within higher education, the erosion of funding and the growth of a *"vicious competitive hierarchy"* between institutions. In addition, the weakening of the student union movement, the *"new managerialism,"* increasing staff workloads and the impact of Teaching Quality Assessment (TQA) and the Research Assessment Exercise (RAE) have increased inefficiency, *"What the Thatcherites intended was to replicate the disciplines of private capitalism: what they achieved was to replicate the waste and irrationality of the Soviet mode of production."*[16] These developments have had a major impact on the management of space and time in higher education and the availability of space and time for academic studies. In the following chapter I move on to focus directly on the steps women students take in order to create space and time for their studies.

Notes

1. Sue Jackson, "Safe Spaces: Women's Choices and Constraints in the Gendered University" (paper presented to Women's Studies Network Conference: Gendered Space, July 1998), 2.

2. Appendix 1c, "Interview Schedule." I draw particularly on question six, where the focus is on women's experience in the college.

3. Jurgen Habermas, *The Theory of Communicative Action. Volume Two—Lifeworld and System: A Critique of Functionalist Reason* (Cambridge: Polity Press, 1981/1987), 325, see chapter 3.

4. Barbara Adam, *Timewatch: The Social Analysis of Time* (Cambridge: Polity Press, 1995), 61.

5. Crescy Cannan, "Enterprise Culture, Professional Socialisation and Social Work Education in Britain," *Critical Social Policy* 42 (Winter 1994/1995): 5–18; Lena Dominelli and Ankie Hoogvelt, "Globalisation and the Technocratization of Social Work," *Critical Social Policy* 16, no. 2 (May 1996): 45–62.

6. Cannan, "Enterprise Culture," 7.

7. Erving Goffman, *Asylums: Essays on the Social Situation of Mental Patients and Other Inmates* (London: Penguin, 1961).

8. Students resisted harassment in various ways, through the existing structures to combat personal harassment, through the students' union, through informal groupings of friends and through other opportunities including external bodies.

9. Jackson, "Safe Spaces," 8.

10. Kelly Coate Bignell, "Building Feminist Praxis Out of Feminist Pedagogy: The Importance of Students' Perspectives," *Women's Studies International Forum* 19, no. 3 (1996): 315–25; Jackson, "Safe Spaces."

11. After Teaching Quality Assessment, the Department was awarded excellence status in recognition of the high quality of approaches to teaching and learning.

12. Kate McKenna, "Subjects of Discourse. Learning the Language that Counts," in *Unsettling Relations: The University as a Site of Feminist Struggles*, edited by Himani Bannerji, Linda Carty, Kari Dehli, Susan Heald and Kate McKenna (Toronto: Women's Press, 1991), 12.

13. Joan Parr, "Theoretical Voices and Women's Own Voices: The Stories of Mature Women Students," in *Feminist Dilemmas in Qualitative Research: Public Knowledge and Private Lives*, edited by Jane Ribbens and Rosalind Edwards (London: Sage, 1998).

14. Heidi S. Mirza, "Black Women in Education: A Collective Movement for Social Change," in *Black British Feminism: A Reader*, edited by Heidi Saffia Mirza (London: Routledge, 1997).

15. Hugo Radice, "From Warwick University Ltd. to British Universities Plc," *Red Pepper* (March 2001, London), 19.

16. Radice, "Warwick University," 20.

10

Women as Centres of Action

" . . . practising centres of action rather than perpetrators of fixed behaviour."[1]

Introduction

IN THIS CHAPTER I discuss the ways in which women students created space and time to study. A wider range of source material is drawn on than in previous chapters. I draw on empirical data from all the research tools used, including the student reflective log[2] where women directly focused on the experience of producing an essay. I discuss women's actions to pursue their academic studies considering their experience external to the higher education institution as well as internal to it. I draw on concepts discussed in chapter 3, in particular the concept of women as *centres of action*.[3] In chapters 7 to 9 I explored the various spatial/temporal frameworks and guidelines shaping women's everyday experience. This analysis provides a basis on which to consider women students as *centres of action*.

I structure accounts of women's experience considering different elements of their action. These elements are drawn directly from the theories of Barbara Adam[4] and Henri Lefebvre[5] in relation to the concept of *rhythm*. I have identified elements of action that focus on movement through both space and time. The chapter is in four parts and explores in turn:

- *Pathways to and through higher education:* where I consider the way women moved through differing spheres of experience.

- *Patterns of spatial use:* where I consider the connections women made between different spheres of experience.
- *Rhythm of studies:* focusing on time, timing and tempo in relation to women's academic studies.
- *Place for studies:* focusing on outcomes, that is, the space that women produced for their academic studies.

This chapter addresses the following research question that emerged from discussions in chapters 4 and 5: "*What actions (physical, mental and emotional) do women of differing social position take in order to integrate higher education into their lives?*"

Pathways to and through Higher Education

Women students constructed routes to and through higher education in the spheres of paid work, home, leisure/community and the college itself.

Pathways through Paid Work

Many women found a pathway to higher education through paid work. Rachel and Maura were encouraged to become students by colleagues at work. Hazel returned to study because of forced unemployment. Experiences of paid work, and contacts there, shaped the expectations and desires of women students with regard to work of the future and the appropriate education needed. For some women, the hatred of routine and physically draining work experiences drove them to seek change through higher education. Rachel felt that paid work had strengthened her academic studies. Knowledge gained through paid work strengthened her understanding of the social world, for example, of social care and trade union structures: "*You learnt by being there every day and doing.*" She used knowledge from paid work as a pathway to knowledge in the academic sphere: "*The knowledge that was already there with some of the subjects we were covering . . . welfare rights, community care, housing. A lot of it was common sense because I'd already been there, within those situations.*" Women connected their current and past paid work experiences to their module and assessment choices in a variety of different ways. Paid work became a subject of study for many students, as dissertation, case study, or autobiographical study.

Those who maintained heavy paid work responsibilities during their studies sought pathways to study through their current paid employment. This involved fitting in certain types of academic work in quiet times at work. Rachel "*used to do it on nights, when everyone had gone to bed . . . If I did an early shift*

I was finished at half two . . . Rest of the day to do stuff. Nights took essays in." Rian said, *"Well actually I was lucky because I worked with somebody until nine o'clock and then it would become very quiet and I'd take my book with me and actually read. There was time I actually sat in the assessment room and did a bit of studying in there which I was able to do and nobody minded . . . If anybody needed me they knew where I was."* Pressure to study during paid work time varied at different times of the academic year and sometimes it was necessary to abandon paid work either permanently or temporarily: *"Revising. I've taken revision to work. I did have difficulty finding the time to proofread one of my assignments. So I took it to work. I parked up and sat in the car and proofread it . . . I've taken my revision to work before today and revised as I've worked and then this year, it was getting close to exams and I was getting a bit panicky so I skived. I did a sicky"* (Susan).

Pathways through Housing and Household

At home, women had to assert their needs as students where expectations of them were quite narrow. Women actively negotiated the social value of higher education with household members, including partners and children. For example, Maura stressed the benefits of higher education to others rather than herself: *"I've tried to be positive with the kids about studying. Because I left school with nothing I want them to go to University . . . So I'm always saying that education is really important . . . 'When mummy finishes this she'll get a job and with the money we'll get a car.' So they see the positive side of it but they also see the negative side of it."* Women enlisted direct practical support from household members including children, parents and partners of both sexes. Of her male partner, Rachel said, *"He'd read through my essays and do the spelling and make sure it read right."* Where possible, women transformed parts of the home into designated study places. There has been a historical demand from women for multifunctional spaces in the home, which would facilitate the multiple tasks they are engaged in.[6] Most of the women students said that the process of studying involved moving through the home and using a variety of spaces for different studying activities:

> But somehow, when I'm doing it on my own sometime I feel a bit isolated there [the dining room]. Sometime I actually go into the lounge, sit down, have the television on and just feel more at peace . . . Doing it there I actually get more done, than actually doing it in the dining room on my own. Sometime I go to the study room where I do a bit of typing. What I do, mainly typing there and other times as well I might feel more comfortable doing a piece of work in the bedroom. It's strange. I sort of move about the house depending on how I'm feeling on that particular day. (Zandra)

Other household members directly or indirectly engaged with this process. Emotions, for example, feelings of isolation, a partner's emotional hostility or a parent's emotional support, opened up or closed down different pathways for different women. Rashida traces her movement for exam revision: "*I worked in the front room 'cause the first week I was upstairs but then I got a sort of depressing feeling so I came downstairs and told my mum I can't work up there and she was OK.*" Irene comments similarly, in relation to work for an essay: "*I used to read up in my bedroom because it was quiet but when I was doing rough work—essay notes I used to do it downstairs in the kitchen. I think it was because I like to have breaks and there was always someone walking in . . . if I didn't understand something I used to talk it through with [my parents] and they used to say what they thought it was and then it used to sort of click in my head.*" Maureen said that if her husband was at home, she was pressured not to study. She did her academic work "*either in the small bedroom, when husband out in the living room on the dining table. At the moment husband goes out to gym with friend for two hours, so I only see him for two hours max/evening during week. However, when home he talks—disrupts me and often argues that work should be done during college time.*" Rian conveyed the complexity and delicacy of negotiations with others over space and time to study. She said that she was reading in a deck chair in the garden. Her male partner started mowing the grass around her and got closer and closer. She asked him did he realise she was working? He said (sarcastically) that it did not look like it. She held up the book she was reading with its very impressive title and he backed off.

For some students it became less and less possible to find a pathway through their housing and household to study (see chapter 8). Some faced outright hostility, felt undermined or were swamped with other things to do. For these, separation was required, which involved both emotional and physical distance. Women employ a variety of strategies to connect and separate the spheres of family and higher education.[7] Physical separation became a pathway to higher education for some women, at least three of whom moved away from partners and into halls of residence on a short or long-term basis: "*Moved into halls. I tried to move last summer. I wanted to split up. Couldn't afford to. Still have to support them financially. Moved beginning of term. Following me, shouting, winding kids up. Absolute nightmare*" (Maura). But separation was a process that also occurred within the home: "*I stopped talking to him about it because it didn't matter what subject it was . . . everything just became personalised*" (Mary). Hazel said she never discussed her academic work with her mother or siblings: "*I didn't really tell family . . . Ripping myself to bits about it. In debate about it . . . Keeping it to myself because I didn't want them to know if I got bad grades."* Hazel felt that her studies would be neither supported nor understood by others in the household. Academic expectations of her from her parents and siblings (and schools) had always been very low. As a consequence she did all her studies in her bedroom unless the house was empty, when she occasionally worked in the kitchen. She never talked about her studies to anybody in the household and did not let them see her studying. She concealed her studying, fearful that their response would deter her from continuing.

Pathways through Leisure/Community

Leisure became the subject of Rian's studies as well as the source of her paid work. As Josephine Burden argues, leisure and sports had become a *"means of taking control and finding meaning"*[8] in her life. Eileen Green[9] argues that the link between enjoyment, relaxation and identity has been understated in academic research. The link between leisure and learning is also significant. Periods of leisure, in the sense of *time for self*, provided pathways to higher education. Relaxation could be an opportunity to review ideas and knowledge: *"Swimming, re-going over what you've put in your essays in your head, exams, memorising the page"* (Irene).

The sphere of community was also significant. Hazel had a mentor at an organisation concerned to further black people's employment opportunities and Kelly *"started using community workers outside college who'd already done the courses as well, and I got a better understanding of what the whole thing was all about then."* Hazel's mother expected her to visit her grandmother in a residential home after she had had a stroke. Although Hazel went on these visits she took academic work with her: *"Sometimes I didn't want to go . . . I would go and I would take some paper with me and make some rough notes. She never really said much."*

When her partner did not recognise the labour involved in her studies, Rian protested. Although she enjoyed her studies she felt it was devalued by being categorised as leisure, *"[He] often sees me reading and watching television for college as leisure. I sometimes think my time spent on studying would have been seen as more worthwhile if I moaned! It seems the more I enjoy college work the more it's seen as my time!"* Women with multiple commitments including child care and paid work felt that the only way to find space and time to study was to completely cut out many leisure activities: *"[Leisure] was things that I did for me, my own time, where I didn't need to tell anyone what I was doing. Could go out a lot more. Had more friends. Since being a student I feel isolated from social life"* (Maureen).

Pathways through the Institution of Higher Education

School is socially recognised as a *highway* to higher education. For some students the transition from school to higher education was non-contentious. However for many students, in a college which recruited from geographically local communities and from a wider age range than is common, the experience of school led to negative expectations. Rachel had *"been dyslexic... school hadn't supported me in that at all."* Susan *"hated school... I don't think you got any encouragement and I didn't do particularly well at school and I knew there wasn't high expectations really... There was a lot of things I wanted to do. I fancied being a window dresser, I fancied being in childcare... They just didn't want to know... so I left."* Even though failed by school, some had positive experiences on access courses. These provided alternative *pathways* to higher education and played a significant role in assisting students to build skills and confidence: *"I started off doing a mature students' course and that was looking at black history and I started doing that through friends as well as particular people I've worked with in the voluntary sector... They were telling me, 'Why don't you use your skills to develop yourself?'"* (Kelly).

Initial advice on an appropriate academic pathway within the college was problematic for three of the younger students interviewed. Two transferred internally and one moved institutions after they had begun their higher education studies: *"I actually initially started on the youth and community route. I came for an interview because I was interested in that aspect but actually I only stayed on that course for a week and I thought, 'Well this is definitely not for me'"* (Suraya). It is not possible to say whether the level of transfer after entry had increased as compared to previous years, but critical questions arise as to whether the heavy marketing of courses in recent years had had an effect.

A recurrent theme in some women's accounts of their entry to higher education was their fear of failure. The marking system was considered central. Higher education was represented as a space where an individual's merit would be measured and quantified in hierarchical ways.[10] For students who failed pieces of work early on in their courses, this fear could be compounded: *"When I actually started my first year I failed two assignments and I thought, 'That's it. I can't do it' and I thought, 'Well, no you can do it'"* (Kelly). Women steered their own routes through the curricula and through particular modules, often selecting topics which connected to the other landscapes of home and paid work that they experienced. Women selected modules with forms of assessment they felt they would achieve well in, usually avoiding exams where possible. There were key exceptions to this. Maura *"actually enjoyed the exams. Vast surge of adrenaline. Sitting there writing, just writing what I want."* Women sought pathways through the tutor and learning support network.

They operated by word of mouth and identified people who they felt would give them the most support. Kelly's academic work began to get higher marks: *"With the help of [tutor] and other support networks within the college . . . I started using [learning support worker] and she put me straight and organised a routine of my work . . . Structured my assignments . . . I started to understand it much better then."*

A major pathway through the college was support from other students. Entering higher education generated changes in friendship networks. One woman (who withheld her name) wrote in the questionnaire, *"Support from friends who are studying but very little understanding from friends pre-college/family members complaining I have become a bore since studying. Sarcastic comments from partner and relatives about the course of study I have chosen."* Christine found *"friends too very supportive, but mostly those going through same or have gone through same that understand best."* Diane described a meeting between her new student friend and her old friends: *"We went to the pub and my friends walked in and it was all 'Oh, hello, how do you do?' and that sort of thing and then she got her roll ups out and their attitude, I couldn't believe it. It just changed and they treated her like shit . . . It was the poverty 'I don't want that nudging in on my life' attitude."* These older friends questioned Diane's choice of course, implying it would lead her nowhere in terms of paid work: *"Is it one of those wishy-washy degrees?"*

Women students gathered in particular groups according to which modules they were studying. In addition they sought out or were pushed towards other students on the basis of social identities for example, in relation to sexual orientation, Clare said, *"It became quite a protective thing between lesbian, gay and bisexual students . . . You had to support each other whether you liked each other or not . . . that was purely because of the identity that people think, assume that you have. Whether you actually agreed with that solidarity just because you were lesbian or gay, you're more or less forced into supporting each other."* Kelly stressed the need for solidarity around colour: *"We had a mixed race chap and he didn't know his colour and it was really difficult at first. But I realised I've got to sit down with him and talk to him and find out his background . . . Through doing that it's helped him to understand himself and know where he's actually coming from . . . it's important."* Diane spoke about the way that mentally distressed women gathered together: *"We didn't say 'Hands up anybody who's got mental health problems,' but we tended to join together. And the smokers. Because we did feel a bit marginalised. But it was generally people who had a mental health problem of one form or another or who had had a struggle in life, whether that could be a child with epilepsy, or a husband who'd been difficult, or a wife . . . We all tended to float together. Same sense of humour."*

Critical Issues: Pathways to and through Higher Education

Although women's different spheres of experience (paid work, home, leisure and higher education) tend to be theorised separately, this research has demonstrated that it is insufficient to focus on the linear pathway from school/home to college when considering women's routes to and through higher education. Women carved out multiple pathways to higher education. Each sphere of everyday activity had the potential to provide such pathways. Each sphere was a potential resource, through providing materials and knowledge relevant to higher education. The motivation to study could be gained from this experience. Paid work, home and leisure were all selected as subjects of academic study. Barriers to academic studies were encountered in each sphere.

Patterns of Spatial Use

Here I explore in more detail the patterns of spatial connection women made in order to pursue higher education. How did women move between the interconnecting landscapes they experienced? The spaces of everyday life were connected; for example, the routines of home were connected to those of work and of college through clock time. I focus on the way that women mediated multiple landscapes in order to pursue higher education. This involved restructuring activities and renegotiating their value: "*Space proposes homologous paths to choose from, while in another sense it invests particular paths with special value.*"[11] Henri Lefebvre argues that when considering *use of space (representational space)* space becomes *alive*, rather than fixed. Rian conveyed the complexity of her movement through interconnecting landscapes. When asked about tensions arising from studying, her response encapsulated the way the landscapes of home, work, leisure and higher education were lived simultaneously but valued differently: "*Having to leave the door. Having to go out the door. 'Cause it's all intermingled isn't it? It's going out to do your leisure. Going out to do your paid work. Being at home for them. It's actually being at home and studying so in a way it was all linked. But, going out to work, in my mind it was always easier. I could justify it more than actually going out to do leisure.*" The act of achieving higher education involved exploiting these connections. For example, Irene carried higher education with her emotionally when she went out to socialise: "*If you're in the house or in your room it doesn't matter if you're studying or not. You don't feel as bad as when you're outside and you haven't got your book with you.*" Diane chose a dissertation topic that connected to her landscape of home: "*It's all about juggling. My mother-in-law. I had to have her cat put to sleep on Wednesday. I know it's an animal. I did my dissertation on companion animals . . . because her husband died two years ago* and everything focused then onto the cat for company." Zandra and Susan developed different strategies to connect the landscapes of household work and higher education. When asked how she fitted housework and studies together, Zandra said she needed to do the housework first in order to feel emotionally prepared to study. On the other hand, Susan did housework when her children were at home and it was less possible to concentrate on studies. She studied when they were out of the house: *"I'd maybe start [housework] till they went to school. When they've gone to school I'd do my college work and then I'd do the tidying up at three o'clock when they came home because I can do that when they're there."* Adopting this strategy meant she sometimes felt emotional unease when having to study in an untidy house: *"Unless it [the mess] got really on my nerves and it was distracting me then I'd have to do it. You're just sat there doing nothing then aren't you? Because I can't do it [study, because the housework] it's bothering me."*

The imperatives of different landscapes, for example their tempo and normative representations, were frequently discordant and made it difficult to make connections. Clare had been on work placement for a year and then returned to the college environment: *"When I came back it was just dead for the first term. There was nothing for us. There'd be an odd lecture here and there. Because we'd got into such a routine of working . . . it was just such a let down. The motivation, everything had gone and . . . between us all we'd all lost that motivation, so we couldn't kick each other on."* She found the intense tempo of her placement, which echoed paid work, discordant with the less regulated tempo of higher education. She had to adjust, *"come back a few paces, go at a slower pace."*

Critical Issues: Patterns of Spatial Use

When considering the ways in which women created space and time to study it became difficult to disconnect one sphere of action from another. Some activities were experienced simultaneously, although valued in different ways. Women, where possible, exploited the connections between different spheres of experience in order to create extra space and time. Different landscapes of experience generated different imperatives for action, for example, the home implied domesticity for many women and paid work was a priority. Women were involved in reinterpreting these imperatives and re-prioritising action in order to make connections with the sphere of higher education.

Rhythm of Studies

"Since time permeates every aspect of existence, it functions as a constant reminder to the physicality of my being, that I am an embodied person inescapably

implicated in my subject matter."[12] The concept of *rhythm* is a vehicle to gain insight into women's action. I now move on to look at the temporal strategies which women used to achieve higher education. What were the multiple rhythms shaping women's experience? How was *others' time* and time for other things dealt with?[13] How was the rhythm of higher education accommodated?

Multiple Rhythms

Individual experience can be conceptualised as a symphony of different rhythms.[14] It was necessary for women to accommodate these in order to achieve higher education. Kelly and Susan convey the complexity involved in this process. Kelly's mother was not well. Kelly and her daughters gave her personal assistance. Kelly's mother in turn looked out for the children whilst Kelly was at college: *"I'm up at half five every morning for the children. Because my mum lives [three miles away] . . . That was catching two buses, or sometimes I caught a taxi and it was like for instance just to get to [college] . . . half five I'd get up. Make sure . . . get their breakfast. Get dressed. Get bathed and everything. Get them sorted out . . . By seven o'clock every morning we were out."* Susan did paid work in the evenings when her partner came home. She gave this account of a typical day whilst a student: *"Up at 7am . . . sort the children out . . . leave . . . half eight . . . drive over to [college] for half nine . . . lectures . . . till half eleven . . . lunchtime . . . one o'clock . . . took the lecture . . . drive . . . to the childminder's . . . back home about quarter to four . . . start tea . . . for five . . . out of the door about quarter to six . . . start work . . . get back home about ten past ten."* For Susan and Kelly the structured clock times of paid work and education (both college and schools) provided the base line rhythm to which other times had to be related. Bodily and household rhythms (eating, washing and relationships) were fitted into the dominant schedule. Multiple rhythms became discordant. Kelly and Rian sometimes found it hard to combine the rhythms of children and higher education. They developed strategies that enabled them to mentally and emotionally switch off: Kelly said: *"If the children are playing up, I've still got to work around them because children always pick on each other, at least mine do . . . At times I just think, 'Just blank it out because [studies are] more important.' I've just got to do it that way. Just get it out of the way, the work. I just blank them off . . . Just do it with them shouting or whatever sometimes."* Rian commented: *"Sometimes when I am reading someone will talk to me [family member] and I will be watching their mouths but I'm not listening."*

The rhythm of Kate's studies were disrupted through her epilepsy: *"Remembering to make sure you take [tablets] 'cause, if you forget, you're worried in case you have another fit . . . Waking up in the morning thinking 'I'm fine.' Getting on with another day. Other times, when I've had them [fits], waking up in the morning and feeling like you've been playing rugby all night . . . It just alters your whole life.*" The rhythm of Qasir's studies were disrupted through her children's illness: "*What I found really hard was when they were ill, because all your instincts are to give up everything and be there for them, and yet you know that you can't, because if you fall behind by a couple of weeks you've had it.*" Attempts to over-regulate the rhythmic complexity of differing demands from work, home, the body and education could break down completely. Diane, speaking of a time before she returned to education, said: "*I had the big nervous breakdown because my job went wrong. Very badly wrong, and I was trying to juggle the . . . job with a home, with a husband, animals, relatives. And I actually couldn't juggle anymore. So when I dropped the balls, bumph, and I just walked out on everything and I just stopped talking, stopped functioning.*" The other effect was that times for intimacy, for self and the body were the first to be disrupted: "*The first thing that went out of the window when I was working and juggling very hard . . . was sex. Our physical relationship broke down totally . . . But as I got stronger at college, as things got better, then each term I got stronger.*" As Henri Lefebvre[15] argues, "*bodies attempt to unite time,*" and this is evident in women's accounts of their experience of accommodating the multiple rhythms of their lives.

Others' Time

Karen Davies has highlighted this as a critical area where inequalities in social power may be identified. Others may coerce part of women's time from them. In paid employment, profit may be extracted from women's labour time. In the household, services may be extracted from women with no return to them personally or to the household as a whole. The benefits of women's time may accrue to others.[16] Women may also be more likely to *resist male time* and take time out of paid work because of their history of commitment to home. In considering time for higher education it is important to address the ways that *others' space and time* in addition to space and time for other things shaped the process. Women wove space and time for higher education out of *others' time*. In addition, time for studies came directly out of time for self, for example from sleep. Kelly spoke of dawn studies: "*I can't work in the night. I work in the morning. Four o'clock to do my studying and if I don't have to come into college I'll spend the whole day from about five o'clock in the morning till about nine o'clock just completely studying, to get my work done. That's how I do it. I can't do it around the children. I can't study at night because I've got so much pressure on me.*" Qasir said, of studying at night, "*After ten, ten thirty. After I'd done everything and then I'd give myself half an hour or so just to wind down and relax and then instead of falling asleep motivate myself to get up and do something.*"

Kate's experience encapsulates the complex negotiations over space and time for studies with other people. Her son resisted her changing his former bedroom (his space) into a study. Her daughter needed personal assistance and could not be carried upstairs. Kate's space and time for studies had to be carved out:

> After [son] left home, I took his fitted wardrobe and all that out which didn't go down too well. He said, "That's my bedroom," I said "Well you don't live here any more." "What happens if I come back?" I said "Tough." I thought, "Sod it. You've gone you've gone!" "Well that's my bedroom." "Well I'm sorry I want to use it. I want my stuff in there" . . . In the end I had to put a little table up [downstairs] because I couldn't work upstairs. [Daughter] was home. If there was little bits I wanted to do on a weekend, I couldn't carry her upstairs because she was too heavy. I couldn't go upstairs and leave her so if I'd work to do at the weekend, which was the main time I could do it, then I had to work downstairs.

Others' time was not necessarily time for tasks, but was also symbolic time. Time when, for example, male partners wanted women to be "in role" as wife or mother: *"I used to wait until [he] was asleep . . . so I'd have to wait an hour and a half, two hours . . . so I could sneak downstairs and do some study for about four hours . . . If I came down he'd be narky all week and it wouldn't be pleasant at all . . . I got into my head 'Sod you, if you don't like it it's tough'"* (Lorna). Obligations on time extended beyond close family. Qasir was expected to provide time and hospitality to extended family and friends and resisted this: *"I had to be really strict. That caused me a lot of problems, but I literally ignored people when they came downstairs and stayed upstairs and if anybody came up I'd actually say, 'I'm sorry, I've only got a few minutes.'"* Although women in halls were often viewed as privileged in terms of time, generally having fewer domestic responsibilities, the impact of collective living meant that *others' time* intruded upon their own. Other students kept Geni awake late at night. She wove her studies around this: *"It sounds like I'm contradicting myself now, but the good thing about them being up till all hours meant that they didn't get up in the morning. So it'd be absolutely silent you know . . . till like two, three in the afternoon, and then the noise would start . . . so I'd just do my work in my room [in the morning] . . . and then go to the computer room."* Higher education for women students was both an individual and a collaborative process. Collaboration took place in a formal sense, through lectures and seminars. It also took place informally through peer, friendship, community and family networks. In this sense time for higher education had a *relational* quality.[17] When students worked together, *others' time* became part of joint space and time to study (see "Place for Studies," below).

The Rhythm of Higher Education

"From a temporal perspective, there is no nature, we constitute nature and we create nature through our actions."[18] There was no single rhythm to higher education. The rhythm of the institution (teaching, assessment) provided a base line to which women tried to adhere. This established periods of lull, flux and intensity, which did not necessarily correspond to other rhythms in their lives. Paid work tended to operate to a more regular beat. Home had irregularity but also a daily and seasonal pattern. Women attempted to integrate the rhythms of higher education with their lives. This, for some women, involved a huge increase in the amount of tasks they undertook, and a reduction in the time to do them. This was more so for women beginning their studies with fewer study skills at the outset: *"I do know from talking throughout . . . with my friends, I definitely put in at least twice as much hours in study time than most of the average students, to keep on top, but that was fine for me, even though I was tired"* (Lorna). The physical energy required in order to sustain paid work, home and higher education was huge for some women and had an impact on their health: *"It's when I stop that's when it starts. I start feeling sick"* (Maura). Time was snatched from every opportunity. Both Maura and Hazel studied on the train to and from college and work: *"Thinking about it when I'm travelling, when I'm on my own. It's a running joke that I don't use the library . . . because I don't have the time for running back and forth"* (Maura).

Time for higher education tends to be represented as time spent reading and writing or sitting in lectures and seminars. However a far wider range of tasks were involved in pursuing higher education, each of which generated their own rhythm. For example, travelling, negotiating buildings, visiting the computer room, queuing, going to the library, carrying books, photocopying, were all actions necessary to achieve higher education. Diane liked *"the library but I couldn't work in the library. I could only just read . . . I'd have to take my books home. I'd have a quick flick through and say 'Bum decision, bum decision but I like this book,' and then I'd return what I didn't want. I was quite systematic about it. Tried not to carry extra crap. If it wasn't good and I couldn't find what I was looking for I'd return it."*

The dominant rhythm of the higher education institution required the submission of assignments at particular times. Preparation required data gathering (through lectures, research, reading) and then deep concentration in order to commence writing an assignment. These rhythmic transitions were reflected in the tempos women adopted for their studies. Such tempos clearly also related to the other demands on their time from paid work and home: *"I used to try and at least read one of the texts . . . so that by the time the assignments came around I'd know a little bit of where to find things . . . try and get* hold of them and look at them . . . the other thing I learnt very early on . . . not to leave it too late . . . otherwise you can't get the books . . . So once the deadlines crept up . . . you just had to put in a few hours" (Qasir). The transition to actually producing (writing) a piece of work required a significant shift in tempo: "Sometimes I felt to myself, 'Well, even though I've got the information there, and everything, how can I start?' . . . You've got too much knowledge in your head to recollect all that information. So I just put it down and go back to it later . . . which is much better because it was clearer in my head then" (Kelly). Women found gaining the momentum needed to actually produce an assignment was not so straightforward. It could not be just switched on: "*Sometime I get an essay. I know it shouldn't be like this but, and I'm there diddle daddling, diddle daddling . . . the final week . . . I have to get a move on . . . then I start panicking and I start working towards the last week or the last two weeks . . . in the beginning I should have started it and somehow I couldn't pick it up*" (Zandra). Some women, by the final year, knew their own essay rhythm perfectly. Kelly took two weeks: "*Say we get assignment, two weeks before I'll do all the research for that particular assignment. That week I'll just get it done . . . I've never had extensions.*" Susan took four weeks: "*Essay deadlines. I usually start well in advance so that I'm not sort of up to the last day. Although this year, with having it quite intense, it sort of got a lot closer than I liked it to get. I like to sort of have a week in hand or something. How long [does an] essay take [to write]? About four weeks for three thousand words . . . from starting reading.*" Rose took two weeks: "*I know it takes a couple of weeks to do an essay. So I leave it till I know my time and I'll start it then depending . . . on how difficult I perceive it to be.*"

A small number of women students wrote essays the night before they were due. This was for a variety of reasons. Some women lacked motivation. Some found it difficult to build up the concentration or momentum required. Some just had far too much happening in their lives. Maura, who did a lot of paid work and child care during the course was involved in a relationship breakdown in her final year. She did her final year essays, including all the reading and research, the night before they were due. Her only advance preparation was thinking about them on the train and getting some books out of the library: "*I write essays the night before basically. I do an all night . . . I don't do anything before. I just get the books out.*" To do this she had to be very focused. She would look at the essay title at an earlier time and then go to the library. She would identify books by using key words in the catalogue system. The night before the essay was due she would use the book indexes to identify quotations. She did not take any notes from books, but wrote the essay straight onto the computer screen drawing quotations from the books around her. She said this was possible because "*I know what I wanted to write in my head.*"

In order to write essays, women students had to sustain a rhythmic complexity that required lots of energy. Once their studies were over, the initial exhilaration was replaced with a feeling of emptiness for some students. They felt uneasy with themselves and their continuing commitments. Higher education had impacted on every landscape of their lives. The rhythm of higher education had interrupted all the rhythms of paid work, home and leisure. The loss of active studying for some women led to feelings of uncertainty and loss:

> It's weird because the last couple of months it's been really intense . . . I've felt I've had no spare time at all. I've felt I've been working all the time. But now, in these past two weeks I don't know what to do with myself. Because I'm just used to working and it's weird when you get home and you've got nothing to do . . . I feel I have to be doing something, you know, and I feel I'm not completely finished and I don't know, it's weird . . . from having no time at all to having so much time now, I forgot how I spent my time. Difficult, I find it difficult to get back into the routine. (Suraya)

> It's funny, just lately [paid work] seems to have got worse. It's that dreading going in sometimes, now that all college has finished. It seems to have sort of like, I don't know, its weird. It's changed since I stopped college. But while I was at college there wasn't a problem there really. I fitted it in. I think I must strive on stress [laughs]. Having days planned. Not having free time. I think on the whole I think I enjoyed it. (Rian)

Critical Issues: The Rhythm of Studies

> This multitude of co-ordinated environmental and internal rhythms gives a dynamic structure to our lives that permeates every level and facet of our existence.[19]

The rhythm of higher education had to be accommodated by women students. The dominant rhythms of everyday life were clock led, relating to the requirements of paid work, school and higher education. Difficulties could arise, in relation to, for example, the more discordant rhythms of children and the body (poor health). If time (as a resource) became problematic in one sphere of action, this had implications for other areas. In order to accommodate higher education, women students carved out (and/or snatched) time from *others' time*, from time for other things, and from *time for self*. The times of other people could interrupt study time (for example, noise in halls of residence) but could also support studies (collaborative time). Studying involved many different forms of action that required complex timing and changes in tempo. The times of studying became integrated into other times and therefore general feelings of unease (or emptiness) could arise when studying ceased.

A Place for Studies

The aim of this section is to identify and explore the places women created for their studies. Studying is a process. I use the term *place* rather than space here in order to identify the results of their endeavours to create space for studies over time. Places are concrete rather than abstract, but they operate at physical, social and emotional levels. Individuals have the capacity to transform the landscapes they inhabit[20] and this is what women students had to achieve. Places to study were shaped from paid work, home, leisure/community and higher education. Such places were often filled with noise and other activities for example, children's homework, family activities, phone calls, caring for others and television. Most women created several different places to study. Virtually all snatched time from different parts of the day. This varied according to different calendars. Rashida got up before dawn during Ramadan in order to eat, and because she was up early she studied: *"Worked on the computer in the early morning 'cause I was fasting then and there was no point in going back to sleep."* The process of studying involved movement. Women created places to study in most of the landscapes they experienced:

> They called it my cerebral area and it was in my favourite room. It was in the kitchen. We put a curtain up. We had one room downstairs, which is part sitting room part kitchen. We put a curtain up and I got a pair of earphones because on the computer I rented, I'd got a CD ROM so I could put my ear phones on and either block everybody out or just listen to music, and I drew the curtain and then I felt I wasn't invading my partner's space. (Diane)

Other people often populated such places. Other activities could be carried out there. Support could be obtained and other people could be included. Kate used to study in the company of her disabled daughter. She realised her daughter was deliberately not asking for assistance whilst she studied. This upset her a lot:

> I can only say I was lucky because [my daughter] was good, as much as, well all right. I could get on with the work better if she wasn't there but when she was there she'd tend to sit and do a little bit of sewing and she looks at her books that sort of thing. The only time . . . if she ran out of wool I'd put another bit of wool in, or she'd got a knot in her wool or something. Wanted a cup of tea or supper, and she'd be trying to talk to me I'd say, "Just give mummy a few minutes and I'll be with you in a minute." "OK mummy," and she'd sew for a bit and then she'd forget and she'd say something else.

Rashida's family helped to create safe places for her studies:

Yeah, 'cause I was really panicking and crying. I just couldn't take it anymore and told my mum that, "I just can't take it anymore. I'm really worried about it." I'd start crying and she says, "Don't worry about it. Look at you, you're not eating. Look at the state of you," and my dad was behind me saying, "You'll pass," and all that, "Don't worry too much right," and all my relatives they all knew about it as well. Everyone knows about it, "Stop worrying. Eat." I wasn't eating anything. I couldn't eat.

Places to study could be created collaboratively. Qasir did her *homework* with the children: *"I actually turned it on its head and said right let's all do our homework. So although obviously I couldn't really work with the three of them, they're butting in every two minutes, I had the pretence of having my hand out to the books and pretending to write, and we've got used to working like that now."* Kelly *taught* her children her studies: *"For instance, if it's to do with . . . women and community, I can make it more exciting by actually reading the book . . . and if it's related to children I will actually relate it to the children as well, and say well 'this particular person's talking about the way children . . . what sort of things they're going through with their parents.'"* Kelly said her daughters learnt to study through modelling themselves on her:

Especially eldest one because it's sort of like reflected back on her now, because to see what I'm doing, she wants to do the same thing as well and I can see that within her. I just hope she keeps it up. I mean her grades are good at school and she's committed and the type of things that she wants to do within community work . . . she's thirteen . . I can see, a lot of it is what I've done . . . It's sort of like stimulated them. Made them look at their work differently. 'Cause they always see me with my books, all the time reading so really . . . they love reading and I can see that within them that it's encouraged them a lot as well.

Zandra often studied with another student, Prabha:

What we would actually do we'd get together, we'd sit round the table. That person will have one book. I will have another. We sort of read and make notes, compared notes. We sometime, when we get a piece, we'd write it down and say "Oh no. That doesn't sound right." We have to reword it . . . but the support was really good . . . I also think working with somebody else makes you . . . I think I covered a lot of work, rather than sometime on my own. More motivated sometime when you're actually working with someone else . . . You can really motivate each other.

Social position shaped the amount of space and time available for studies. For example, poor heating, low income and the need to do paid work left women with less time to study. Women were involved in considerable negotiations, at paid work with employers and colleagues, at home with families and other co-residents, at leisure with friends and family. In order to study effectively women had to find a place where their studies were considered legitimate. This involved negotiating the social value of higher education. As space is coded in relation to social division[21] this involved women in re-conceptualising space and time as legitimate to study in. Diane had to convince her parents, employer and friends of her right and need to study: "*Mum wanted to clean around me! Parents wanting me to stop working and talk to them . . . social commitments already made had to be broken . . . Had to cut down hours [paid work]. Boss not happy . . . had to work when nobody was in house.*" Rian faced pressure from a partner—"*What, no tea again? If you'd planned your time better we wouldn't have to stay at home at weekends*"—and from friends—"*I haven't phoned you as I thought it was your turn to ring me.*"

Clare's experiences of trying to study were challenged directly and indirectly because of homophobia. Whilst she constantly asserted her right to a place in higher education she eventually came to the conclusion that it might have been better to be more cautious about coming out as lesbian in the higher education environment. She had not expected the level of hostility she encountered as a student. When asked what advice she would give to other women in the same position who wanted to study she said, "*Go ahead. Do it. But caution I've got, especially around sexuality. It's portrayed as an OK place. It's sink or swim. It's a battle. Be aware that if you're going to be out, then everybody knows and you've really got to deal with that . . . You shouldn't have to be in that position about whether you're out or not.*"

In order to create places to study, as well as restructuring their everyday lives, many women had to reconceptualise their everyday roles. Qasir came to see her role as a mother as an important space for higher education that included her family: "*For my children, I will always remember . . . [the work of Paulo Freire] and try and get hold of one or two books and give them a more positive view of being black. And again on gender issues, it's given me a wider perspective. I mean globalisation and the economy. I've never really thought economically before. The word 'economics' just scared the hell out of me.*" Susan identified the ways higher education had strengthened her mothering. She had learnt the importance of building the educational confidence of her own children through her own academic struggles: "*I look at things from other people's perspective and it's definitely helped me with my children to help them at school. The confidence to help them and the importance, like . . . to encourage them and develop their confidence and if they say they're thick or something like that I go balmy . . . I say you are not, you're very clever . . . and it works, I'm sure it works.*"

Arlie Hochschild[22] argues that an examination of emotion enables the researcher to map areas of conflict, complexity and paradox in social life. The intensity of demands on women students' time, particularly when essays were due in, generated feelings of guilt, stress and anger. In addition, for some women self-esteem was an issue: *"Being a mature student, self-esteem and confidence in what you're writing is sometimes a strain"* (Diane). Women tried to actively create a physical environment that was restful and calming as a place for studies, for example by playing music, tidying up, and lighting candles. For some women, higher education had been identified as *not their time*, because of age, class or community for example. Kate came into higher education after her male partner left home. She had gone through feelings of despair prior to becoming a student and saw higher education as her lifeline. However, at first she felt she was too old to study. Her feelings shifted: *"It's something that's been there all these years that I've never been able to do. I know for a fact that if he'd still been at home I'd never have done it. I actually feel that I've actually done something for myself . . . for me just for me."* An emotional place to study appeared to be a place that allowed a balance between space and time for self and for others, whether family or community. Both Lorna and Kelly re-valued themselves, justifying their right to higher education and highlighting their potential social contribution. Lorna expressed this as selfishness: *"I feel as if I'm more selfish really. We're brought up to think doing anything for yourself . . . is a selfish thing . . . but I don't believe it is . . . so I stand up for myself . . . I make sure I have time for myself . . . instead of giving all my time for other people. I distribute it out for other people and for myself so that's making me feel a whole better person."* Kelly expressed this as emotional strength: *"It's made me stronger . . . When I think about it, why didn't I start years ago? Probably I wasn't ready for all that. To go through the experience what I've gone through now. So it's made me aware of certain things that I can give back within the community. I've realised that I've got a lot of skills that I can give. I can help people with. I've just got to do something about it."* The role of emotion in creating space and time to study cannot be underestimated. Emotions have the capacity to reflect, map and begin to redress inequalities in power. Both Maura and Hazel had difficult lives and faced oppression for very different reasons. Both had to contend with emotional pain. For both of them higher education represented a pathway out of these circumstances. Maura said that studying *"creates space where you can explore a lot of things . . . I absolutely loved it. It's been a real challenge for me."* Hazel said, *"Studying was the making of me. I still feel I don't want to give it up."* Susan came to see that the emotional hurdles she had gone through had also been learning resources: *"I would say that all obstacles that you come across are all learning experiences and that it all comes in and . . . when you look back . . . Well I know it was difficult at times and caused tears and upset . . . but its definitely a learning experience and to look at it positively."*

Critical Issues: Places for Studies

Higher education was a process, part of which involved women students in creating places to study. Such places were constructed in each sphere of activity. They were often shared with people involved in other activities. They could also be places for collaboration. Women had to re-conceptualise the places of everyday life as places to study. In order to create places to study, some women had to re-value themselves and justify their higher education as time well spent.

Conclusion

Women students' higher education action occurred at different levels and in different spheres. At the physical level it involved, for example, rearranging rooms. At the social level—negotiating the right to study with partners. At the emotional level—resolving feelings of guilt about the effect of studying on their children. Women made connections across spheres of experience in order to include higher education. Such action involved social interaction and personal reflection (as well as actual studies). I have focused on women students as *centres of action* and conceptualised their action as embedded in their everyday life-worlds, which are different according to social position, personal history, stage in the life-cycle and so forth. These life-worlds do not determine the activities that women students will engage in but form a basis for and a field of their actions: *"Is not social space always and simultaneously both a field of action . . . and a basis of action?"*[23] Higher education is a process. It involves far more than mental engagement. Studying, in this case, was accommodated across each sphere of women's daily lives and into a wide network of relationships. Particular pathways to higher education arose or were obstructed in different ways for different women. The times of paid work, the household, children, leisure and friendship were all relevant to the process of studying and studying was combined with these times.

"Spatial practice tends to confine time, and simultaneously to diminish living rhythms by defining them in terms of the rationalised localised gestures of divided labours."[24] Social responses to women students were associated with the particular roles they were expected to fulfil within the routines of paid work, leisure/community and domestic labour. Their engagement with higher education could result in entrenchment of certain attitudes, for example, the view that studying and mothering could not be combined. In order to accommodate higher education, women had to carve out pathways through physical and social space and time and manage the complex tempos of life in new ways. In the final chapter I review the research, considering the key issues that

emerge from the analysis in chapters 7 to 10. I discuss the effectiveness of space/time as research tools. In addition, I include a reflexive analysis of the researcher log and discuss the impact of the researcher's action.

Notes

1. Barbara Adam, *Time and Social Theory* (Cambridge: Polity Press, 1990), 74.
2. Appendix 1b.
3. Adam, *Time, Theory*.
4. Adam, *Time, Theory*.
5. Henri Lefebvre, *The Production of Space* (Oxford: Blackwell, 1991), 207.
6. Cynthia Rocke, Susanne Torre and Gwendolyn Wright, quote from Appleton's Journal (1870), in "The Appropriation of the House: Changes in House Design and Concepts of Domesticity," in *New Space for Women*, edited by Gerda R. Wekerle, Rebecca Peterson and David Morley (Boulder, CO: Westview Press, 1980).
7. Rosalind Edwards, *Mature Women Students: Separating or Connecting Family and Education* (London: Taylor and Francis, 1993).
8. Josephine Burden, "Leisure, Change and Social Capital. Making the Personal Political" (paper presented at Leisure Studies Association International Conference "The Big Ghetto," July 1998), 4.
9. Eileen Green, "Women Doing Friendship: An Analysis of Women's Leisure as a Site of Identity Construction, Empowerment and Resistance," *Leisure Studies* 17 (1998): 171–85, 176.
10. Kelly Coate Bignell, "Building Feminist Praxis Out of Feminist Pedagogy: The Importance of Students' Perspectives," *Women's Studies International Forum* 19, no. 3 (1996): 315–25; Sue Jackson, "Safe Spaces: Women's Choices and Constraints in the Gendered University" (paper presented to Women's Studies Network Conference, Gendered Space, July 1998).
11. Lefebvre, *Production, Space*, 42.
12. Barbara Adam, *Timescapes of Modernity: The Environment and Invisible Hazards* (London: Routledge, 1998), 7.
13. Karen Davies, *Women and Time: The Weaving of the Strands of Everyday Life* (Aldershot: Avebury, 1990).
14. Adam, *Time, Theory*, 174.
15. Lefebvre, *Production, Space*, 203.
16. Davies, *Women, Time*, 204–6.
17. Davies, *Women, Time*.
18. Adam, *Timescapes, Modernity*, 13.
19. Adam, *Timescapes, Modernity*, 13.
20. David Crouch and Jan te Kloeze, "Camping and Caravanning and the Place of Cultural Meaning in Leisure Practices," in *Leisure, Time and Space: Meaning and Values in People's Lives*, edited by Sheila Scraton (Eastbourne, U.K.: Leisure Studies Publications, 1998), 54.

21. Linda McDowell, *Capital Culture. Gender at Work in the City* (Oxford: Blackwell, 1997), 12.

22. Arlie Hochschild, "The Sociology of Emotion as a Way of Seeing," in *Emotions in Social Life: Critical Themes and Contemporary Issues*, edited by Gillian Bendelow and Simon J. Williams (London: Routledge, 1998).

23. Lefebvre, *Production, Space*, 191.

24. Lefebvre, *Production, Space*, 408.

11

Conclusion: Women Creating Space and Time

"As time, timing, tempo and temporality we can recognise some of the complexity of that which is ultimately individable... Through its rhythmicity life becomes predictable. Thus the focus on time helps us to see the invisible."[1]

Introduction

MY APPROACH HAS been especially concerned to widen the boundaries of research into women's everyday experience. I have examined higher education as an aspect of this.[2] I have not restricted the focus to their engagements with the institution of higher education or with their homes. Women's experiences and actions in one sphere were closely connected to their experience and action in other spheres. Throughout I have argued that wider social forces do not determine individual action but mediate the form that action takes. Below, I discuss the principal findings from each area of analysis and draw further connections. I draw connections between different spheres of women's experience and between different theoretical conceptualisations of space and time as *practice, representation* and *action*. I also draw connections between my own experience as a researcher and the experience of women students. This is achieved through reflection on the log that I kept to record my own thoughts, experiences and feelings as the research progressed (see chapter 6). This final chapter is in five parts, considering:

- higher education: restricted ground;
- higher education: ambiguous meaning;

- women and higher education: regaining ground and meaning;
- researching women's lives through space and time;
- researcher's reflective log.

Higher Education: Restricted Ground

In chapters 7 and 9 I considered the dominant *spatial temporal practices* (the frameworks for action) women experienced in the spheres of home, paid work, leisure/community and higher education. These spheres, I argued, were the true boundaries of women's experience of higher education. The spaces women occupied during the day had emerged from processes over space and time. Each place women occupied in their everyday lives, whether paid work, home, higher education or leisure, also had a local spatial and social history.[3] Women's places of paid work were predominantly external to their homes. The home was generally conceptualised as a private place, concerned with domesticity, recuperation from paid work, and care.

The radical restructuring of the public sector in recent years, initiated by the New Right and pursued by New Labour in the U.K. had a direct impact on the ways in which women students experienced space and time. They were studying in an era of national social policy change that affected every sphere of their experience. Cuts in public expenditure coupled with the deregulation of formerly state-led services meant that most spheres of experience had certain characteristics. These were less direct centralised control, the devolution of areas of decision making, new managerial techniques and also an increased level of monitoring of performance, which amounted in some cases to an increased level of surveillance of behaviour. These women students, for example, worked in a particular niche of the labour market that had arisen through the spatial/temporal restructuring of work in recent years. This restructuring had involved the deregulation and privatisation of state run services and the drive for short-term profits and cost effectiveness. Women students normally worked outside traditional working times, fitting in paid work in the evenings, at weekends and during the night, for example. A major source of employment was private care work, which had increased as a result of policies concerned with rolling back the state and stimulating independent care provision. These women experienced pressures to work in cost-effective ways, whether they were based in industry, cleaning, shop work or social care. All the agencies they worked in had developed a competitive ethos. Mother students were highly involved in both paid work, and unpaid work in the home.

Cuts in public expenditure and the deregulation of state welfare impacted on women students in different ways. For example, cuts in social security benefits for students increased the pressure on women to engage in paid employment. Cuts in student grants and the introduction of loans increased the pressure to generate further income in whatever ways they could. The introduction of the "right to buy" (public housing) and reduction of publicly owned housing stock meant that women students either lived in owner occupation (with the attendant mortgage commitments), stayed in the parental home, or lived in the high rented private sector, halls of residence and housing associations. Lack of capital investment in student accommodation meant that many women experienced halls of residence as very poor quality housing.

In relation to the sphere of home, the impact of the above restructuring, deregulation and cuts in state support were twofold. Women were under pressure to generate more income themselves in order to contribute to their own higher education and household finances. If they lived with other family members, both younger and older women felt under pressure to create time in the household for others to work, through, for example, taking more domestic responsibilities. They sometimes faced greater scrutiny of their actions by other household members and found they had to justify the time they spent in studying. It would be fair to say that the increasing financial pressures on households exacerbated this situation. In addition, women worried about the increased surveillance they were under from the state, in relation to claiming welfare benefits and child support payments and by local education authorities and the Students' Loans Company. They were under increased pressure to raise income to finance their own higher education, but the ways in which this could be legally achieved were limited, because of cuts in entitlement. They were under greater scrutiny in the higher education institution in relation to attendance. The college was expected to report lengthy absenteeism to student financing bodies. Women were loath to disclose details about personal income in the research. One woman said she had deliberately understated the amount of paid work she had done because of fear of the Benefits Office reclaiming her welfare benefits. It may well have been the case that other students had also done this and that findings about paid work in the research therefore underestimate the amount of work women students actually do. An ongoing concern for lone parent women was that the Child Support Agency would contact former (sometimes violent) partners if they tried to claim benefits.[4] During the period of the research, to my knowledge, at least one person in a student's household was prosecuted for social security fraud. The person concerned was claiming benefit as a lone parent and failed to declare that she cohabited with a student. Vindictively and quite inappropriately, she was imprisoned even though this was a first offence.

In chapters 5 and 9 I argued that the deregulation of higher education had increased competitive practice both between higher education institutions and within individual institutions. This directly affected women's experience of creating sufficient space and time to study. From the very outset of their contact with higher education institutions the competitive ethos prevailed. For example, some may have been given poor advice about courses. They shifted to other courses early in their academic careers. Could they any longer be certain that the advice they received about courses was truly impartial because of the pressures on tutors to maximise student numbers? Competitive practice in higher education also impinged on staff security. For example, the pressure to carry out research and to publish decreased the amount of daily contact between some staff and students. Staff security had also been weakened by the emergence of differential terms and conditions and the increased use of part-time contracts.[5] More recently appointed staff found their pay was lower than that of staff that had been in post longer, even though the expectations of them were the same. Many members of staff had little time to develop a research profile because of heavy teaching loads. Student contact with practising researchers was limited as a result. As with other sectors, the reduction of centralised control in higher education, coupled with spending cutbacks and the development of a competitive ethos were matched by increased forms of surveillance of educational practice in the institution. Different forms of monitoring continually evolved ranging from student-led and peer-led evaluation of courses, to inspection by management and external institutions such as the Quality Assessment Agency. Administrative loads and stress levels had increased amongst staff. Students on the one hand felt they had some voice in curriculum development, teaching and learning issues. As a rule they were included in course committees, for example. On the other hand they suffered the effects of increased student numbers, semesterisation, and limited access to staff at crucial times, for example when assignments were due.

The impact of the restructuring of *spatial/temporal practices* directly impinged on the space and time women had available for higher education. Higher education had become restricted ground in the political drive to save public expenditure, to increase productivity and to reduce the role of central government in direct service provision. Even though the expectation under New Labour was that access should increase and that in particular students from low-income households should be encouraged to enter higher education, the material conditions which would allow students to study in effective ways (space, time and money, for example) had been dramatically reduced. As a result, women students faced considerable hurdles in finding space, time and resources to study. These hurdles were greater than those faced by students some years ago, when both grants and welfare benefits were available to those on low incomes. The ground for higher education in the lives of women students involved in the research was restricted in a variety of ways. Space and time for studies had to be snatched from space and time for paid work, home and leisure/community.

Higher Education: Ambiguous Meaning

In chapters 8 and 9 I discussed the dominant *spatial/temporal representations* (guidelines for action) women said they experienced in their everyday lives. From these accounts higher education emerged as an ambiguous space, open to highly contested interpretations by women students and those they interacted with. Space and time for higher education had to be negotiated with other people both in and outside the household as well as in the college itself. Women often had to justify their engagement with higher education to themselves. I explored women's experience of the different *landscapes* of home, paid work, leisure/community and higher education. Four areas emerge as critical: the significance of financial transactions; hierarchical values associated with space and time; normative assumptions regarding the use of space and time; associated emotional dissonance.

The Significance of Financial Transactions

Jurgen Habermas[6] argues that money is a major motor of system integration in advanced capitalist systems. E. P. Thompson[7] identifies the nature of money as a measure of *time spent*. As with other forms of system integration, money may be experienced as a neutral artefact, but it carries immense real and symbolic power (see chapter 3). For the women I interviewed, whether money passed hands was a way of measuring how highly time spent on certain activities was socially and individually valued. The ways in which money passed hands and the amounts involved were a measure of the power of certain individuals and groups, and their ability to extract profit from the labour of others. Money provided access to goods, services and higher education. It was a key medium of power and social regulation.

Work outside of the domestic sphere was normally paid for. Negotiating the value of time spent in paid work was a major concern. Relative to other activities, paid work had high social value amongst women. Engagement with even very low paid work was central to New Labour rhetoric during the period of the research. They stressed the responsibilities of welfare recipients to engage in paid work wherever possible. Some types of paid work were valued more highly by women students and they aimed, through higher education, to increase their paid work status. For example, permanent work was generally preferred to temporary; high paid to low paid; technical, managerial and administrative to manual and support work. Patterns of paid work (including hopes for future work) were gendered, with the emphasis on social care related occupations.[8]

In the home, money passed hands in less formal ways than in paid work. Decision making over household income was a key indicator of where power in the household lay.[9] The passing over or withholding of money was a means of regulating behaviour. Parents and partners could withhold money from women who deviated from expected roles. Women themselves could use the little money they had to negotiate extra space and time for studies, through for example, paying for childcare or paying children to do housework, even where this conflicted with their feelings. Household labour was generally unpaid and negotiating the value to the household of such work was very difficult. Most of the older women were continuing with domestic responsibilities they had had for several years. This work may have been previously invisible to others in their households but when they became students its visibility increased. The pressures on the whole household of accommodating the financial costs of a woman's higher education were complex. The household had to decide who should bear the additional costs of higher education. What income could they afford to lose in order to create the space and time for the woman to study? Where could additional income be raised from and who should raise it? Integral to this process was the need to justify higher education as a legitimate drain on household resources. Was higher education a justifiable long-term investment? For many households the answer was ambiguous and uncertain. Karen Davies[10] has argued that some women from working class backgrounds have low expectations of further and higher education as a source of financial returns. However, women who identified as working class in this study had seen that the potential gains were in stark contrast to their past manual employment experience. Members of their households sometimes saw older women students as a risky investment. A student with past learning difficulties had low expectations of success, which were shared by her family. A woman who identified as middle class faced critical questions about the likelihood of the particular courses she was on leading to good paid work (see chapter 10). The majority of women students perceived higher education as one of their only routes to greater financial security in the long term. Many said that they had come into higher education because they saw it as a route to better pay and more autonomy for themselves and their households. The financial implications of women's engagements with higher education led to intense negotiations around what higher education involved financially for the household and what the long-term financial outcomes would be. There were no straightforward answers to these questions.

Hierarchical Values Associated with Space and Time

The close connection between money and the way time is conceptualised (as for example, time wasted) involved a hierarchy of social values. Time spent in paid work was considered a clear priority. Those engaged in paid work were more likely to be considered entitled to decision making in the household and to leisure time. Time spent in domesticity was also generally highly valued for women, even where they themselves sometimes perceived it as time wasted. The physical space of home was associated with domesticity for women. Questions therefore arose as to what were legitimate ways of spending time whilst present in the home. Some women found that they were expected to revert to the domestic whilst at home. When they wanted to study, negotiations over space and time could be intense and filled with emotion. Leisure time was generally conceived of as something to be earned. Other activities took precedence and only when these were completed was leisure justified. Women who pursued leisure often felt guilty whilst doing so. If a woman had not completed an assignment that was due in, she felt her leisure was undeserved. Certain leisure activities, leisure spaces and the concept of leisure itself were valued in different ways by different women. Leisure for some was an extension of mothering or of obligations to other kin. The sphere of leisure overlapped with the sphere of community as women's household work extended into the community. For both younger and older women, leisure was a sphere for building new relationships and friendships. But leisure remained ambiguous, viewed as secondary to other spheres, yet a central part of everyday activities and involving intentions to overcome oppressive experiences.

The meanings ascribed by women and their households to higher education were closely connected with their conceptualisations of paid work, home and leisure. In the research the ways in which higher education was interpreted shifted in relation to different spheres of experience. Male partners might imply that the legitimate space for higher education was the institution itself. Some of them conceptualised this as the "work site" of higher education and saw no reason for studies to be brought home. As Rosalind Edwards found,[11] some male partners in particular disliked the clutter of studies. A woman's social position was critical in whether she got support with her studies and the form that this took. Older women and women who were mothers faced particular obstacles and challenges to their studies. Mother students in particular were often considered to be neglecting other duties.

Conceptualisations of higher education were closely related to those of paid work. Higher education was conceived of as a route to better paid work and was itself considered as a form of work, which involved a labour and production process.[12] But the connections between higher education and paid work were not straightforward. Higher education lacked the pace and tempo of some paid work settings. Where women gained pleasure from higher education they might conceptualise it as a relief from work. Partners might conceptualise higher education as women's *time for self* and not as legitimate work time. Finding space and time for higher education for some women was akin to finding leisure time, in particular where it was conceived of as *time for self* or *selfish time*. Women had to justify their engagement with higher education and in some cases felt the need to disguise the pleasure they gained from it. Although higher education was sometimes conceptualised as diverting women from domesticity, some women came to conceptualise higher education as an extension of domesticity. They argued it fulfilled their obligations to their children and households and that it was a route to greater financial security for the whole household. Being engaged in higher education created an environment where both children's and partners' learning could also be enhanced. Hence higher education was conceptualised in ways closely connected to the meanings of women students' other daily involvement with paid work, home and leisure/community. It was conceptualised in relation to those spheres and as a route of transition in all those spheres. The meaning of higher education was ambiguous and complex for these women as the meaning shifted in relation to their struggle to legitimate their studies.

Normative Assumptions Regarding the Use of Space and Time

Closely connected to the above conceptualisations were a set of normative assumptions about the use of space and time. Studies interrupted the normative practices of paid work, home and leisure/community. For older women, studying challenged normative expectations. Studying was generally associated with childhood and youth. Paid work was a space to provide labour to others in exchange for money, and generally not associated with studies (unless the paid work involved training or was directly linked to academic studies). However, many women challenged these assumptions in their use of resources from paid work to study. They snatched space and time from paid work in order to study. They made their paid work the focus of their studies. Similarly, although leisure was assumed to be space and time for relaxation and pleasure, many women used leisure spaces to strengthen their studies, through friendship, time for themselves and relaxation. Studying was sometimes likened to leisure. The higher education institution itself was a space to pursue learning. However, the space of the college functioned in different ways for women students. In addition to learning, it was referred to as a social and leisure space, a noisy space where studies could not be pursued and a space for paid work (cleaning and in the canteen). It was a living space (for women in halls) and a space for domesticity and relationships. Most of the spaces women encountered were designed for specific purposes. These were represented in their design, in the layout of rooms and in the artefacts contained. When women wanted to study, the spaces they occupied could create physical barriers to this. For example, the layout of living space in the home and halls of residence meant quiet space was hard to find. Women students challenged normative assumptions regarding spatial use and created multi-functional spaces.[13] The meaning of time spent in higher education was laid out in the shape and organisation of the institutions occupied as they moved through the day (institutional time).

Associated Emotional Dissonance

Creating space and time to study and justifying this led to *emotional dissonance*.[14] For example, some women felt uneasy in the college itself, either because they felt they were not like *real* students (because of age or self perceived ability) or because their right to study there was challenged (because of racism or homophobia). In the spaces of paid work, women were challenged because of their student status. In their homes, some mother students felt guilty about studying. Did it reduce the quality of their mothering? The range of feelings women experienced about their studies, both inside and outside the college, exposed the particular barriers that they faced. Their emotions were also a means of responding to and of overcoming such barriers.

Women and Higher Education: Regaining Ground and Meaning

The concepts of *centres of action* revealed the tactics and strategies women adopted. I found that the routes into higher education that women took were complex and overlapping. Their experiences in paid work, school, access courses, the home and leisure/community all provided pathways. For example, knowledge from and expectations of paid work shaped both the desire to study and curriculum choice. The community provided pathways, from links with voluntary organisations, learning support groups and peers. Students however, faced barriers to their progress in all these spheres. Women carried knowledge of one sphere of experience to their engagements in another. What spatial connections did women make to achieve higher education? Their negotiations in relation to higher education were embedded in their everyday experiences. They negotiated the value of higher education in relation to the dominant meanings ascribed to different spheres (work, household and leisure/community). They physically negotiated space and time for higher education out of other activities. They reordered and re-conceptualised activities.

For example, they combined household and higher education, paid work and higher education, leisure/community and higher education wherever possible and when motivated. At a physical level this involved reordering quantitative space and time. Socially it involved redefining the meaning of action in relation to the hierarchy of values and normative assumptions attached to space and time. Emotionally it meant justifying higher education to themselves and creating safe and peaceful environments that were conducive to their studies.

The *rhythm* of their studies was fascinating. As with Barbara Adam,[15] I found that the dominant *clock times* (and institutional time) of paid work, school and higher education regulated their days. The other rhythms of the household, their bodies and to some extent the wider environment interrupted these dominant rhythms. Where the regular tempo of life was interrupted, through ill health for example, this caused emotional dissonance for women. Their own times (particularly their own body times, health and sexuality) were sometimes the first to be neglected. I found that women students in their final year of studies usually had a very detailed time plan in relation to their studies, whether they did essays the night before they were due in or the month before. Women generally knew what length of time a piece of academic work would take them and acted accordingly. The processes involved in higher education studies were not just about reading, attending classes, writing and revising. I found that higher education involved numerous activities in every sphere including paid work, home and leisure/community. In the sphere of the higher education institution it involved many time consuming tasks, including movement through buildings, queuing for and carrying books, learning and using technology and collaborating in friendship and peer groups. Women said it required an immense amount of physical and emotional energy to gain the momentum to actually write assignments. They found commencing writing difficult. At this stage of academic production they often had to *switch off* other commitments. Other aspects of the higher education process could usually be accommodated with less difficulty.

Women were involved in trying to comprehend what higher education meant in order to justify their engagement with it in every sphere. They were involved in manipulating and reordering other activities to find space and time for studies. They carved out space and time from *others' space and time* and created different places to study. The places they produced in order to study arose from the framework and guidelines set out for them[16] and out of the dialectical relations of space and time. The *landscapes* they moved through were transformable to create space for higher education at a physical, social and emotional level.

Researching Women's Lives through Space and Time

In chapter 2 I reviewed feminist theoretical perspectives and argued that research is strengthened through approaches that draw on both materialist and post-structural analysis. I argued that a *critical realist* approach facilitates this and emphasised four principles underpinning such an approach (see chapter 2, "Feminist Critical Realism"). I will now evaluate the research in the light of these principles. This section is in three parts considering the following questions: What has been added to research about women students' experience of higher education? Have central regimes of power been identified in women's experience without generalising the experiences of different women? How effective were the spatial/temporal concepts drawn on in comparison to other possible approaches?

What Has Been Added to Research about Women's Experience of Higher Education?

Viv Anderson and Jean Gardner[17] argue for a deeper understanding of the ways in which women's everyday experience connects with the higher education process. In my research I have shown the ways in which the actual practice of higher education by women students is embedded in their everyday experience. It is important to acknowledge that it is at the level of daily events where pathways to higher education are created and barriers to learning are confronted.[18] I have shown that it is not only social attitudes that prevent certain groups of students from achieving. A whole range of practices that they encounter, both within the college and external to it, transform the nature of their studies. Madeleine Gromet[19] refers to the dialectical relation between home and higher education and argues that these two spheres influence each other in the practice of higher education. This research extends her analysis to include the dimensions of leisure/community and paid work. I argue that these spheres are also highly significant in the ways that individual students practise higher education. There is a *symphony of rhythms* shaping women's lives and it is in relation to all of these rhythms that space and time for higher education is produced by women.[20] As Chris Heward points out, examining women's experience of higher education in the context of a *linear occupational career* fails to adequately convey this process.[21] This research has exposed the ways that a group of different women created space and time for higher education. It provides vivid detail in relation to the complex of processes and actions involved. Studying is theorised as movement through space and time, involving a wide range of interaction, adjustment, restructuring and reconceptualisation. The methodology included research tools and concepts that drew attention to the different ways space and time are experienced and shape social reality.

Have Central Regimes of Power Been Identified in Women's Experience without Generalising the Experiences of Different Women?

On the one hand, the use of the concepts of space and time enabled attention to be drawn outwards from women's accounts of their experiences. These accounts in themselves could not explain, but add insight to the wider regimes of power giving rise to them.[22] Space and time as *practice* drew attention to the specific material positions of women students. For example, the restructuring of higher education, state welfare and paid work became more visible. Space and time as *representation* drew attention to the dominant meanings contained in the spaces women occupied. Central regimes of power became visible as they shaped individual perceptions and feelings.

It was not possible to generalise too widely about women's experience because spatial/temporal concepts highlight differences in position, experience and personal history. Attention to women's emotions in the research acted as a gauge of where they personally felt unsettled by normative expectations. Age, mother status, residence, colour, religion, geographical heritage, class and sexuality intersected differently with dominant normative expectations. Individual women's feelings and experiences arose in relation to their specific position and could not be generalised to other women. Dominant *spatial/temporal representations and practices* gave rise to *emotional dissonance* for women differently.

How Effective Were the Spatial/Temporal Concepts Drawn on in Comparison to Other Possible Approaches?

In this research I drew on broad conceptualisations of space and time as I felt that they would do the most justice to women students' accounts of their experience. Some of the conceptualisations discussed in chapter 3 were not followed through. For example, I did not develop a detailed typology (listing the many characteristics) of space and time along the lines of Anthony Giddens[23] or Georges Gurvitch.[24] I felt such a typology would be as likely to constrain the research analysis as to facilitate it. The attempt to impose a hypothetical typology on the data would limit the findings. Another area of concepts that I did not include in the research is from psychoanalysis. I found the focus on the relation between mental process and the social world, although sophisticated, too narrow for my purposes. Conceptualisations of time from psychoanalytic theory were not broad enough to explore the ways in which the spatial and temporal framework and guidelines impacted on personal action at a number of levels (see Carol Watts, who shares this view).[25] In this section I discuss the strengths of the conceptualisations I drew on.

The concept of *spatial/temporal practices* drew attention to contemporary social arrangements in relation to paid work, home and leisure/community and highlighted the routes women took in order to include higher education in their lives. This concept enabled lines to be directly drawn from women's experience at the micro level, to the wider relationships of power and institutional arrangements that provided a framework for their everyday lives and their academic studies.[26] Higher education was conceptualised as a practice that extended into the other spheres of everyday life, as well as having its own institutional base.

The concept of *spatial/temporal representations* drew attention to the different levels at which space and time functioned in women's lives (physical, social and emotional) or *"the different depth levels of social reality."*[27] The knowledge, social value and meaning associated with different spaces provided guidelines for action. Women who were making change in their lives, through entering higher education, had to steer these guidelines. They had to adapt to and ignore them when necessary if the messages they received conflicted with their need to study. They had to restructure, negotiate and reconceptualise space and time in order to create space and time for higher education.

The concept *centres of action* gave visibility to the wide range of factors influencing action. *Reason* has been a primary focus in social action theory. Cohen[28] points to the limitations of this approach, as relationships of power become less visible (see chapter 3). In this research, rather than focusing solely on mental processes, the emphasis is on *movement* and *rhythm* in space and time. These concepts maintain the visibility of the macro level of social reality (the framework and guidelines for action) as well as the multiple rhythms shaping women's everyday lives. The concept of *being* in space and time underpins the concept of action. The account I give of women students' action in chapter 10 is consequently detailed, holistic and realistic. It also encompasses similarities and differences between women.

I have argued that space carries social messages and provides social functions. I would argue that to focus on the spatial does not direct our gaze away from the social, but demonstrates the complex ways in which social power operates. In focusing solely on social phenomena one may lose sense of the dynamics of social change. Social phenomena may be conceptualised in static and determinist ways. Having regard to the spatial maintains the visibility of the dynamics of social change. It is hard to say what distinguishes the spatial/temporal concepts I adopted from other sociological concepts. There is clearly considerable overlap between, for example, the *spatial* and the *social*, between *representational space* and the concept of *culture*, and between *spatial practice* and the concept of *social networks*. *Centres of action* may be yet another term for the concepts of *agency* or *praxis*. I would argue that the critical difference between the spatial/temporal concepts I have drawn on and other similar sociological concepts that do not directly refer to space and time is in their power as analytical tools. For instance, the concept of *spatial/temporal practice* draws attention not only to *social networks*, but also to dominant regimes of practice, such as segregation and separation, public/private zones and the influence of *others' time* on personal time. *Spatial/temporal representation* draws attention to the social power contained in space and place. The concept of *culture* also has this potential but, as Nira Yuval Davis[29] has argued, *culture* is often mistakenly conceptualised as a fixed realm. The concept *centres of action* is different from either that of *agency* or *praxis* (which in turn differ from each other). *Agency* conveys a sense of intentional action whether based on emotion or reason. *Praxis* describes a moment of social change. *Centres of action*, on the other hand, conveys an image of multiple actions and draws attention to the wide range of social and personal rhythms framing those actions. The use of such a concept avoids accounts of action as, for example, solely determined by the economic, biological or social events.

The power of spatial/temporal concepts lies in their facility as research tools. They direct the gaze to a range of detail that enriches research. If space and time are made visible then a variety of other issues also maintain visibility. These include for example: the context of events; difference in social position; the connections between different spheres of experience; processes of exclusion and separation; the relationship between the past, present and future in events and the relationship between knowledge of the world and material conditions.

Researcher's Reflective Log

In this final section I reflect on the research by analysing the ways in which my own knowledge and experience of the world shaped it. I analyse the reflective research log (see chapter 6) from October 1998 to June 1999, when the main part of the data gathering and analysis took place, and where I kept notes of my experiences during the research. I consider myself as a *centre of action* engaged in research. What common ground did I share with the women students whom I studied? How did I create space and time for the research and how did this affect the findings? I address the final research question, "*What are the implications of the researcher's actions for the findings?*"

Pathways of Education (Paid Work)

As discussed in chapter 3, the spatial/temporal relations of paid work provided a context in which the research emerged. As a lecturer in higher education, my work had been subject to considerable external restructuring in recent years. Firstly, cuts in student finance made it difficult for the college where I worked to maintain student recruitment at the high levels that had been reached. There was pressure on staff to design and market new programmes of study to attract more students. Secondly, the emphasis on vocationally oriented programmes had led to increased emphasis on the work related placements.[30] Thirdly, the traditional tasks of teaching and marking had increased. Semesterisation had led to an increase of modules, which tended to be shorter and more intensive. The consequence was an increasing squeeze on time available for both students and staff as is evident in my log: "*7/5/99 My time squeeze: new degree and MA programmes to be written . . . [validations]; placements; marking; seeing desperate students about their time to study [trying to extract essays].*" In addition, the necessity of having a research profile became all the more important as the Research Assessment Exercise shaped staff appointments in the universities.

The professional role of academics is complex and contradictory. They provide pathways through the higher education institution for students. They have the power to close off access to higher education if students' work (and behaviour) is not of a standard acceptable to the institution and to employers (the latter in the case of vocational courses). They have the power to facilitate student progress through effective teaching and learning support strategies. The complexity of the professional role is manifest in different higher education practices, for example, exam boards, marking, tutorials and teaching. There is a two-way dependency between students and staff. Students are dependent on staff to find pathways through higher education. Staff rely on students as both sources of employment and of reputation. The complexities of this relationship are evident in these three extracts from my log, which reflected the range of regulatory higher education practices I was involved in: "*2/2/99 A student came to question her mark—very nicely and apologetically but also upset . . . my power is enormous—I forget—felt self-critical; 15/2/99 Difficult work day . . . a student placement breakdown . . . very difficult to be fair all round; 10/3/99 Lots of great chats at work with students over past couple of weeks . . . It's great to talk to students about what they want to talk about which is not content of essays but how the heck to do them.*" Why did I choose to focus on women students' space and time to study, consequently choosing to focus on barriers and routes to learning from outside the higher education institution as well as within? Reflecting on my research so far has led me to realise that I chose to focus on aspects of my own professional role concerned with the facilitation of student learning. Although not positing women students as a *problem* for the institution, I focused on their everyday lives and did not examine the practices of the institution of higher education in as much detail. In order to truly strengthen women students' position it may have been more legitimate and radical to focus directly on the forms of regulation they were subject to within the higher education institution. It was through a complex of rules, regulations and normative expectations that women students pursued higher education. My own professional role as a paid worker clearly influenced the direction the research took.

Pathways of Education (Student)

I myself was involved in higher education relations as a registered student. I was dependent on the academic institution where my studies were based, and on my academic supervisors to facilitate my pathway to a doctorate. Research expectations came through the practices of that institution, for example those related to the supervision, registration, approval and progression of research students. In terms of personal supervision, I identified a woman supervisor who I felt would help me find a pathway through the institution and through the research. My choice was made because of my prior knowledge of both her academic interests and her supportive approach and a hope that she would legitimate my approach (as feminist and sociological). This proved the case and my confidence grew. Nevertheless, I depended on the regular support and approval of my supervisors to an extent I had not realised until reading my log. Research decisions I made, particularly at the early stages of the research were directly affected. My need for approval and affirmation reflected the power embedded in our respective roles as supervisor and research student: *"6/10/98 Still worried about questionnaire . . . can't give it out till [supervisor's] commented; 23/11/98 Was really glad [supervisor] thought feminist theory paper OK; 14/12/98 [Supervisor] boosted my confidence; 14/1/99 . . . bit on tenterhooks till I hear from (supervisor)."* The *"behind scenes of power and control"*[31] for me as a research student came through the expectations and interests of other academics I had contact with in relation to my research.

Pathways of Education (as Knowledge)

The bodies of knowledge I drew on (feminist, sociology of space and time, social divisions and women students) led me backwards, through references in the texts I read, to the earlier sociology of social action theory. Reading Beverley Skeggs[32] led me to the work of Pierre Bourdieu[33] and reading John Urry[34] to the work of Henri Lefebvre.[35] Journeying back through the sociological literature, I began to understand more the foundations of the concepts I was dealing with. However, the path I took meant that particular writers were very difficult to understand at the time that I read them, in particular because I did not have the foundation knowledge they were drawing on. There is a tradition within academic culture that involves a belief in *expert knowledge*[36] that is a belief that it is possible to master a particular academic discipline in its entirety. The advent of multidisciplinary approaches makes this far less possible. Clearly, there is a limit to the amount of academic knowledge a student is able to consume.[37] The process of engaging with knowledge through texts in the way described above, that is *backwards* in time, led me to feel a loss of connection with both feminist ideas and with women students themselves even though feminist writers and women students had provided pathways to this earlier writing. The urgency to read on created difficulties: "*14/1/99 . . . must read more.*" The concepts of space and time began to lose specificity as different meanings and interpretations were presented in the literature, which conflicted with the specific and concrete meanings which women students had given them: "*15/2/99 Time/space is now everything and therefore in danger of disappearing again into generalisations.*" The process of the literature review sometimes felt mechanistic: "*22/1/99 I'm playing the paper game in three ways—getting theory to fit; finding theory to build; shuffling bits of paper around with weak headings.*" Although my attachment to the theoretical concepts of *space, time* and *gender* grew out of my own experience and pilot interviews, there remained a tension between ideas from the sociological literature and the knowledge of women. Skeggs[38] sums up this tension succinctly: "*My desire for control of knowledge initially led me to produce representations which were more consistent with this desire than with the experiences of women.*" It is for these reasons that feminists have emphasised the need for collaboration in research and the importance of allowing reflexivity into the research process. In reflecting on these aspects of the research, I can now see the way that dominant bodies of knowledge shaped the process.[39] It is extremely difficult to make a radical *break* with existing epistemologies. In fact methods which bring existing knowledge together rather than attempting to reject them may in fact allow for more progress.

Pathways of Home

In considering the pathways of home I have pursued in relation to the research I will discuss two aspects. Firstly, my past memories of pursuing higher education at home. Secondly, this research and its relation with home. It is only at this stage that I have clearly recognised the impact of my home-based experiences in the past in formulating the research. Family ill health and male violence affected my childhood. My early schooling was important to me because of the stability it gave me. During my first degree (1970–1974) I had a baby and my male partner of that time was not supportive. My first postgraduate studies (1975) were shaped by the break-up of that relationship and by domestic violence. By the 1980s I lived alone with my older son. My MA studies (1982–1984) I recollect as a more pleasurable experience where space and time to study were more in my control. There is no doubt that these earlier experiences of home and higher education shaped the focus of my research.

More recent home-based experiences were significant in a different way in shaping the research. I had emotional support to study from my current partner. I had chosen part-time paid work because of mothering responsibilities (my younger son).[40] I felt guilty and tired in full-time work and had sufficient money to manage in part-time work. As a consequence of these choices, my experiences of studying in the domestic sphere were similar to those of other mother students I interviewed. The hierarchy of values attached to time in the household was such that time for paid work took precedence over (unpaid) study time. As with other women students, my studies in the home involved movement, change and negotiation over space, time and equipment: "*26/10/99 Waiting to use the computer. Two hours before I pick up [my son].*" I chose to study where I felt comfortable and those choices related to the meanings and feelings associated with different parts of the house. I sometimes felt inhibited using the study. Being surrounded by books and paper was not relaxing and made me feel hemmed in: "*14/1/99 Today didn't want to use the study because like to move round house.*"

When I became ill, studying was suspended for six months because of the pain. When I recommenced studying, this was in small stages and I lacked the energy to deal with complex ideas. I also found it difficult to sit, so writing at the computer was problematic. On reflection, my choice of research topic was directly related to my landscape of home. It reflected the wider divisions of space into public and private zones where paid and unpaid labour is experienced. It problematised home and household as a critical dimension of higher education. This focus was directly grounded in my own past and ongoing experience throughout the research.

Pathways of Leisure

Women's formal engagement with leisure is often associated with household responsibilities rather than personal felt needs. There were brief references to formal leisure activities in my log and little other comment. My explicit account of leisure referred to family-based activities alone. However, periods of leisure, in the sense of time for self and relaxation also provided a pathway to the research. Relaxation became an opportunity to review ideas and knowledge. In reflecting on my own log, I can see how even periods of sleep were sometimes productive: *"14/1/99 Woke up thinking 'Is higher education production or consumption from student's point of view?'"* Leisure as community and friendship was also an opportunity to talk about my work as well as run from it: *"19/7/99 I'm finding this really hard. Analysis of interviews . . . doesn't help because of stomach-ache. Abandon and go see friend."* I borrowed academic books from mothers who were lecturers and who were waiting in the school playground when I picked up my son. One other mother who was a student read and commented on an early paper I had written. I talked with other mothers at school about the problems of combining paid work and mothering. I became involved in community work to develop child care services and secure more time for both paid work and studies. All these activities may be conceptualised as related to my research.

The pleasure gained from the research and the positive emotions associated with studying meant that for me, the research in itself sometimes became a pathway to a feeling of leisure, for example, *"22/10/99 . . . overall enjoying it."* Since my illness the research also afforded some relaxation from physical pain and therefore some relative leisure. Small amounts of studying distracted me from the pain. Research became akin to leisure as I had been housebound for several months and the research took me away from *home* in my mind.

Patterns of Spatial Use

I now consider the complexity of the interconnecting landscapes in my life as a researcher and my patterns of spatial use. In constructing the research, the connections between the spheres of home, leisure and education had to be mediated. Rosalind Edwards[41] has argued that many women's attempts to connect family and higher education are thwarted. Being successful often necessitates separating family and higher education. But whatever choice women are ultimately driven to, the landscapes of paid work, home and studying are experienced simultaneously. This extract from my log shows the way feelings about home, relationships and paid work came together in the process of research. I felt low in relation to the research and my paid work: *"22/10/99 Long gap in log because I've been getting down to it. Have been looking at 'action.' Is literature review taking me backwards or forwards? Also my progress paper sounds very pretentious. Feel isolated. Work: my job share has gone up to full time. Home: partner very busy all the time, [son happy]. Research: internal, introverted."* Issues in one sphere had direct implications for another. I will now discuss three patterns of spatial connection I made during the research.

Firstly, my movement between paid work and the research. Secondly, my movement between home and the research and thirdly, my movement between different bodies of knowledge.

Connecting Landscapes of Work and Study

The complexity of my professional role as a lecturer was discussed above. On reflection, my paid work role was expressed in the research methods I actually adopted, rather than those I constructed which are outlined in chapter 6. My place of work generated research questions and data both formally and informally. For example, I chose to interview a woman student I met in the corridor at work, who was not part of the original sample I had identified: "*12/6/99 [Hazel] . . . hadn't been going to interview but talked in corridor. Thirst for education. Carer. Dreadful educational start. Most moving of interviews.*" My research continually fed my teaching. I introduced students I was teaching to bodies of knowledge and methods that I had come to through my own research. For example, I encouraged a woman student doing an assignment related to community care to draw on the study of women and time by Karen Davies[42]: "*23/11/99 [Susan is] reading Davies and applying to her life for an essay.*" I encouraged another student to draw on Allison Jaggar's[43] work on emotion and feminist epistemology, in order to explore issues related to the experience of women as carers at home: "*10/3/99 . . . chats with student about 'outlaw emotions'—feeling and caring.*" When carrying out research interviews my dual role continually shaped the approach I took. Research decisions were affected by my responsibilities as a tutor. My approach to one student was affected by her disappointment with a mark I had previously given her. I had hoped to interview her for the research, but "*12/6/99 I was hesitant because [Geni had been] disappointed with . . . mark [I'd given her]. 'Don't let that stop you. Show respect for her views in interview—may compensate for mark.'*" I avoided probing too deeply into personal issues in research interviews. I did not want to affect students' academic confidence: "*12/6/99 Didn't want to upset her/unsettle [Maura]. Very able—high pressure; 12/6/99 In interview I didn't probe [Geni] . . . happy student, good relationships I perceived. 'Don't rock boat.' We can learn from emotions of pleasure as well as pain.*" When I asked one young woman from the selected sample to come for an interview, I saw the look of disappointment and worry on her friend's face. I therefore asked them both to come together. This was a complete break from the methods I had chosen (individual interviews and questionnaires). I was ill prepared to conduct a joint interview. Nevertheless, it was necessary to do this because students associate meetings with tutors as opportunities to strengthen confidence and get clarification on studies. It was important that as their tutor I ensured students felt they were fairly treated even though the research interview was not connected to their own studies. They did not necessarily understand this because all their previous encounters with me were as a tutor: "*12/6/99 [Suraya and Rashida] together. Asked [Rashida] in canteen. [Suraya] looked worried [still awaiting marks]. Said 'come too.' Very effective. Oldest sister and youngest sister from different families. Reflected on each other's experience.*"

In chapter 8 I discussed the ways women students collaborated with each other. Throughout the research I found this attitude extended to me. The women I was interviewing helped me explore some of the issues in detail. Sometimes I chose to directly share research questions with students although I had not intended to. There were particular areas where I needed more insight and women respondents actively collaborated with me. They appeared to empathise because they too had been conducting research for undergraduate dissertations. For example, Rose helped me to look at life in halls of residence: "*12/6/99 . . . co-operative discussion to help me get to bottom of halls issue.*" Irene explained in detail her experiences working in a residential home for older people, and we discussed the impact on young women: "*12/6/99 . . . co-operative discussion. Carers.*" My roles as tutor, researcher and student informed the research on an ongoing basis.

Connecting Landscapes of Home and Research

Above I discussed the pathways I took through home to the research. On reflection, my hypotheses about women students and higher education were grounded in my own home-based experience. My experiences of home also shaped the construction of the research on an ongoing basis. My landscape of home informed the questionnaire and interview design and areas I probed and held back from in interview. Overall, conversation about home in the interviews was richer and took more time than other areas. Perhaps this was because students held back more from critiquing the institution of higher education itself because I was their tutor, but it is possible I steered the discussion towards home and probed more in that area.

Connecting Landscapes of Knowledge

Above I discussed some aspects of the pathways through knowledge that I took and the tension between different epistemologies. These two extracts from my log demonstrate this tension: "*11/10/98 I'm grappling with feminist theory paper. Keep getting an exciting framework and then not sure I grasp the ideas—particular meanings of specific terms, e.g. structuralist/post-structuralist. Think I've got it then it disappears like smoke. Also fail to see connection with* what I'm doing but feel it's stuff I've got to get out of the way; 14/12/98 Got to fit feminist research with dead white male [and some feminist] theory now." Uppermost was the tension between research theory, people and texts. I was guilty of the classic academic illusion that focusing on personal experience and women students' experience was a subjective process yet focusing on texts was not: *"16/11/98 Main problem—feel strong need to avoid subjectivity. Need to identify relations giving rise to the research and also evaluate the research process. These aims pull me in different directions. Identify relations? I turn to books. Evaluate the process? I turn to self. Must do both. Use self as pathway to reading?"* The belief that texts provided a more objective and reliable knowledge of the world informed my practice. On reflection however, my choice of books was subjective and the knowledge contained in the texts I read represented particular perspectives. Although I felt I was reading critically and evaluating arguments as I read, in practice I was gathering knowledge from different areas and bringing it together: *"23/11/98 My approach is to read loads and then drag it kicking and screaming together and this probably involves reinventing it all together."* The epistemological landscape I moved through was laid out for me as represented in the academic disciplines of sociology, geography and feminist studies. I sought ways to connect these perspectives, following pathways indicated by specific authors.

The Rhythm of the Research

Pat Whaley[44] writes about the complexity of research time: *"We operate in different time zones and our body clocks have great difficulty responding—there is the personal, the political, the financial . . . Furthermore, research time is quality time. I find that while some aspects of research can be done fairly mechanically, spasmodically, whenever and irrespective of mind, mood or motivation, much of research is just not like that."* In relation to home, analysis of my log shows the ways that time for research was carved out of others' time and time for other things: *"25/8/99 [My son's] poorly . . . I've written recently about body times interfering with external regulation! Well here it is. I've been sitting . . . ploughing through tasks . . . notes, cards, references; money for [fees]; book to order; chucking and sorting. Midway went to the GP . . . my son vomited outside the health centre."* The rhythm of my own paid work as a lecturer imposed deadlines on the research, for example my involvement with a degree validation and teaching: *"6/4/99 Time is definitely shaping the work. I've got three and a half days and I want to deal with questionnaires i.e. develop for analysis . . . I want a framework to conceptualise material linked to concepts for example, others' time, public/private, spheres of action—tall order. I've got degree validation in two weeks. Have just stopped reading press cuttings [for teaching]."* The higher education calendar for students imposed deadlines on when the data gathering could take place. For example, when would students be available to fill in questionnaires? When should interviews be conducted? I felt it was important this was after the final exams: *"14/12/98 Feeling quite calm and organised. I've given out questionnaire A to all students now and everyone was really great . . . with filling it in—though some hiccups and a couple of questions . . . too wordy and confusing, and income [questions]—student worries about disclosure of [information about] income? Though overall lots of positive responses!"*

The research also had an internal rhythm and calendar that stemmed from orthodox ideas about the process research is expected to take. My acceptance of this orthodoxy can be seen in the ways I laid out my research plans in the initial research proposal document.[45] The timetable I proposed suggested a logical order of stages to the research that was difficult to adhere to. It provided an internal *drive* to the research: *"27/1/99 Have now completed . . . theory paper [very dense] and higher education paper [simplistic]. Need to plough through methods reading and plan ready for analysis . . . Must just plough on. Hope I'm getting somewhere; 15/2/99 Now on research methodology paper. At first bored and boring. Then 'eureka' . . . now struggling again [institution]."* My personal energy and bodily needs also informed the process: *"22/1/99 . . . running out of writing steam; 23/11/99 I'm speedy today—too much coffee."* My back injury resulted in suspension of the timetable and a general slowing down when I recommenced studying. My personal satisfaction with the research was greater when I managed to accommodate the complex time demands in my life and to conform to the various calendars that regulated the work: *"14/12/99 Lots of hard work but I feel quite creative."* The act of research was experienced as an act of production. I, as a researcher, felt pride in the *dominant timescapes* I experienced in relation to the research.[46]

Research time also had a social value. Certain activities were identified as legitimate research activities and others were not. Reading my log I see that I made minor references to reality that I considered external to the research. I discussed the connections between home, paid work and research because that was the focus of my research questions, but I ignored other events and commented only when these generated very strong emotion that invaded the research. For example I made this aside in the log during the war in Serbia and Kosovo: *"6/4/99 . . . And there's dreadful genocide taking place."* I did not evaluate such events in terms of the research although on reflection they clearly impacted on both my own and the respondent's experience at the time. These external events were conceptualised as disconnected from the research.

Addressing the multiple rhythms shaping the research and the social value accorded to time use, reveals the way that research emerges in particular ways, giving visibility to some phenomena and not to other. Emphasis on the outcomes of research encourages researchers to present the research process as in their control from the outset. For instance, when writing up, they may present the research questions and methodology as fixed from the outset when this is not the case: *"22/1/99 I'm starting on the research methods paper. This is the most difficult yet because I've already started the research."* An examination of process in social science research shows the way *"behind scenes of power and control"* shaped the questions asked and the focus of study on an ongoing basis.

A Place for Research

Physically, the landscapes of paid work, of higher education, of home and of leisure/community all provided places for the research. The research was in a sense locked into these landscapes and the temporal rationales, calendars and associated rhythms. Research questions and the methodology were generated from the interconnections between these spheres. Priority was given to some relations over others. Socially, the categories of knowledge I drew on came from dominant epistemologies associated with the spheres of home and household, paid work, leisure/community and higher education. I tried to find a social place for the research through mediating the differing concepts from the literature and from women students. My own social beliefs shaped that process. At the emotional level, in retrospect I can see how the research was driven in part by my past experience of higher education as a woman. Emotional memories reverberated in the research in the way I attempted to articulate women students' dilemmas. I was clearly also seeking to re-articulate personal dilemmas I had faced as a woman student. I was very concerned to find a political place for the research. The account I have given of its construction could be viewed as very determinist. I have stressed some of the external processes shaping the research. But clearly I was also active in making choices which I believed would further the interests of women students. Although the *"behind scenes of power and control"* may have worked against such an outcome, my acknowledgement of those relations was critical. I would support Beverley Skeggs[47] who discusses the difficulty and the importance of maintaining a political/advocacy perspective in research:

> In the micropolitics of the research process, power relations (generated through positions in institutional and disciplinary locations) are occupied, yet they may not be recognised or valued. But if they are, if I have the power to authorise the accounts of these women, what does it mean to them? I do not have the power as a feminist researcher to convert their cultural capital into symbolic capital or give it an economic value in order for social and economic change to occur. What I can do is challenge those who have the power to legitimate partial ac-

counts as if representative of the whole of knowledge and to challenge classification systems which position "others" as fixed.

My research was the outcome of a wide range of actions on my part, some conscious and others not. The research was also the outcome of a network of other action, by for example, the writers I referred to, the institutions I practised from, and the individuals I interacted with during the process. By explicitly using concepts of space and time in order to analyse the research process I give visibility to the importance of these events in the construction of the research. The use of space and time as a framework for the reflexive analysis forces me to attend in detail to events which were highly influential but which may have remained invisible in more traditional approaches. For example, both my own autobiography and the research protocol in the university I attended become more visible.

Concluding Remarks

The approach I have adopted to researching space, time and gender enriches understanding of the complex processes involved when women try to change their lives. By drawing together spheres of experience which are often theorised as disconnected it has been possible to develop a holistic account of women's actions and achievements. By ensuring that space and time remain visible in women's lives it has been possible to gain insight into both the everyday routes and pathways women take and also the ways that major systems of inequality shape experience. I have considered emotion as a critical gauge of inequalities and as a means of achieving change. I have applied the concept *centres of action* to women students and this has proved an important vehicle for understanding the specific ways in which they achieve higher education. In addition, through analysing the spatial/temporal relations shaping the research as it progressed, I have drawn attention to the influence of researcher action on the findings.

This research provides a foundation for several areas of further enquiry. In relation to higher education, firstly, there is a need for further detailed research into contemporary *spatial/temporal practices* of higher education and the impact of these for different groups of students, for example, the effects of finance, time-tabling, semesterisation and specific programmes of assessment. Secondly, further research is needed into the *spatial/temporal representations* of higher education. What does higher education mean to different groups of students in relation to their everyday lives? What learning support is appropriate if they are to carve out pathways to higher education and overcome the everyday barriers to learning that they encounter both within the academic institution and external to it? Research is necessary into the ways that the structures of learning support within the colleges and universities might include the understandings, for example, of those living in halls of residence, of mother students, of young students and of students with different class and geographical heritage. Thirdly, it would be fruitful to apply the concept of *centres of action* to study the lives of particular groups of men and women of different backgrounds pursuing higher education. The important relationship between higher education and other life transitions (relationship, sexuality, leisure/community, religion, housing and paid work) would thus become more visible.

The potential of the conceptualisations of gender space and time discussed here have far wider application than higher education. The concepts and methodology developed give visibility to different levels of reality, for example, complex and intersecting relations of power (industrial, economic, political) the links between the macro and micro in daily experience, the commonalities and differences in women's experience related to social divisions and social position and above all the tactics and strategies developed by women to carve out space and time in their own interests.

Notes

1. Barbara Adam, *Time and Social Theory* (Cambridge: Polity Press, 1990), 169.

2. Dorothy Smith, *The Everyday World as Problematic: A Feminist Sociology* (Milton Keynes: Open University Press, 1987).

3. Paul Bagguley, Jane Mark-Lawson, Dan Shapiro, John Urry, Sylvia Walby and Alan Warde, *Restructuring: Place, Class and Gender* (London: Sage, 1990).

4. Fran Bennett and Marilyn Howard, *Child Support: Issues for the Future* (London: Child Poverty Action Group, 1997).

5. *Times Higher* 3/9/2000.

6. Jurgen Habermas, *The Theory of Communicative Action. Volume Two—Lifeworld and System: A Critique of Functionalist Reason* (Cambridge: Polity Press, 1981/1987), 171.

7. Karen Davies, *Women and Time: The Weaving of the Strands of Everyday Life* (Aldershot: Avebury, 1990), 26, citing Thompson.

8. This clearly linked to the academic routes these women had chosen which were principally related to youth, community, health and social care.

9. Jan Pahl, *Money and Marriage* (Basingstoke: Macmillan Education, 1989).

10. Davies, *Women, Time*, 199.

11. Rosalind Edwards, *Mature Women Students: Separating or Connecting Family and Education* (London: Taylor and Francis, 1993).

12. Liz Stanley, "Feminist Praxis and the Academic Mode of Production," in *Feminist Praxis*, edited by Liz Stanley (London: Routledge, 1990).

13. Davina Cooper, "Regard between Strangers: Diversity, Equality and the Reconstruction of Public Space," *Critical Social Policy* 18, no. 4 (November 1998), 465–92.

14. Arlie R. Hochschild, "The Emotional Geography of Work and Family Life," in *Gender Relations in Public and Private: New Research Perspectives*, edited by Lydia Morris and E. Stina Lyon (London: Macmillan, 1996).

15. Barbara Adam, *Timescapes of Modernity: The Environment and Invisible Hazards* (London: Routledge, 1998), 9.

16. Henri Lefebvre, *The Production of Space* (Oxford: Blackwell, 1991).

17. Viv Anderson and Jean Gardner, "Continuing Education in the Universities: The Old, The New and The Future," in *Educating Rita and Her Sisters: Women and Continuing Education*, edited by Roseanne Benn, Jane Elliot and Pat Whaley (Leicester: National Institute of Adult Continuing Education, 1998).

18. Michel Foucault, "Truth and Power," in *Power/Knowledge: Selected Interviews and Other Writings, 1972–1977*, edited by Colin Gordon (New York: Pantheon Books, 1980a), 114.

19. Madeleine Gromet, "Conception, Contradiction and Curriculum," in *The Education Feminism Reader. Part III: Knowledge, Curriculum and Institutional Arrangements*, edited by Lynda Stone (London: Routledge, 1994), 150.

20. Adam, *Time, Theory*, 74.

21. Chris Heward, "Women and Careers in Higher Education: What Is the Problem?" in *Breaking Boundaries: Women in Higher Education*, edited by Louise Morley and Val Walsh (London: Taylor and Francis, 1996).

22. Smith, *Everyday World*; Liz Kelly, Sheila Burton and Linda Regan, "Researching Women's Lives or Studying Women's Oppression? Reflections on What Constitutes Feminist Research," in *Researching Women's Lives from a Feminist Perspective*, edited by Mary Maynard and June Purvis (London: Taylor and Francis, 1994); Sherry Gorelick "Contradictions of Feminist Methodology," in *Race, Class and Gender: Common Bonds, Different Voices*, edited by Esther Ngan-Ling Chow, E. Doris Wilkinson and Maxine B. Zinn (London: Sage, 1996).

23. Anthony Giddens, "Structuration Theory: Past, Present and Future," in *Giddens' Theory of Structuration: A Critical Appreciation*, edited by Christopher Bryant and David Jary (London: Routledge, 1991).

24. Georges Gurvitch, "The Problem of Time," in *The Sociology of Time*, edited by John Hassard (London: Macmillan, 1990).

25. Carol Watts, "Time and the Working Mother: Kristeva's 'Women's Time' Revisited," *Radical Philosophy*, no. 91 (September/October 1998): 6–17, 15. In her analysis of Kristeva's essay "Women's Time," she comes to a similar conclusion. She critiques Kristeva's work with reference to the work of Thompson and Lefebvre and argues that the psychoanalytic approach, although powerfully conveying the way that time is experienced in relation to social power, does not sufficiently address the temporal framing of experience: "*A series of differentials simultaneously and multiply lived in everyday life (at the workplace and home, and in the home as workplace); marked by the passage of the day and the seasons; according to the task, the employer, the technological means available; as representing various degrees of autonomy and imposition.*"

26. Sherry Gorelick, "Contradictions of Feminist Methodology," in *Race, Class and Gender: Common Bonds, Different Voices*, edited by Esther Ngan-Ling Chow, E. Doris Wilkinson and Maxine B. Zinn (London: Sage, 1996).

27. Gurvitch, "Problem, Time," 7.

28. Ira J. Cohen, "Theories of Action and Praxis," in *The Blackwell Companion to Social Theory*, edited by Bryan S. Turner (Oxford: Blackwell, 1996).

29. Nira Yuval Davis, *Gender and Nation* (London: Sage, 1997), 67.

30. Although this was not a new feature of the programmes I taught on, there was increased emphasis on the quality of supervision of student placements.

31. Derek Layder, *New Strategies in Social Research* (Cambridge: Polity Press, 1993).

32. Beverley Skeggs, *Formations of Class and Gender: Becoming Respectable* (London: Sage, 1997).

33. Bourdieu, Chamboredon and Passeron, *Craft, Sociology*.

34. John Urry, "Sociology, Time."

35. Lefebvre, *Production, Space*.

36. Elliot Jaques, "The Enigma of Time," in *The Sociology of Time*, edited by John Hassard (London: Macmillan, 1982/1990), 25. He warns students about attempting to draw on the sociology of time without fully grounding themselves in the history of ideas.

37. In addition, the Research Assessment Exercise has triggered an excessive plethora of journals and other published material that makes the task of reviewing relevant literature even more difficult.

38. Skeggs, *Class, Gender*, 38.

39. This echoes Janice Moulton, "A Paradigm of Philosophy: The Adversary Method," in *Women, Knowledge and Reality, Explorations in Feminist Philosophy*, edited by Ann Garry and Marilyn Pearsall (London: Routledge, 1996).

40. Rosalind Edwards and Simon Duncan, "Supporting the Family: Lone Mothers, Paid Work and the Underclass Debate," *Critical Social Policy* 53, no. 17 (November 1997), 29–49. They argue that such choices emerge in resolving dilemmas about paid work and mothering through specific *"gendered moral rationalities,"* differing values are attached to mothering and paid work by women situated differently.

41. Rosalind Edwards, *Mature Women Students: Separating or Connecting Family and Education* (London: Taylor and Francis, 1993).

42. Davies, *Women, Time*.

43. Allison M. Jaggar, "Love and Knowledge: Emotion in Feminist Epistemology," in *Women, Knowledge and Reality: Explorations in Feminist Philosophy*, edited by Ann Garry and Marilyn Pearsall (London: Routledge, 1996).

44. Pat Whaley, "Women Researching in Continuing Education: Voices, Visibility and Visions," in *Educating Rita and Her Sisters: Women and Continuing Education*, edited by Roseanne Benn, Jane Elliot and Pat Whaley (Leicester: National Institute of Adult Continuing Education, 1998) 97.

45. Appendix 3.

46. Adam, *Timescapes, Modernity*, 5.

47. Skeggs, *Class, Gender*, 37.

Appendix 1
Research Tools

Appendix 1a: Semi-Structured Questionnaire

Would you help with some research into women students' experiences of fitting studying into their lives? Last year ____ and I spoke to several women students after they had completed their degrees. The major issues they raised related to the tensions, difficulties and arrangements involved with securing sufficient time and space to study because of their many other commitments.

I hope to develop a fuller picture of women students' experiences and commitments over the coming year and am asking all women in their final year of studies at ____ to fill in the attached questionnaire. I will be giving out a much shorter questionnaire later in the year and hope to interview some of you after the course has ended.

All responses will be treated with confidence and if I draw on the questionnaires I will change any identifying detail.

This research will not affect your assessment in any way.

Dot Moss.

Name

Course and year of entry

Date of birth

1. **Paid employment**
1.1 Have you engaged in paid work whilst a student at ____? Please delete: Y/N
 If no please go to 1.4.
1.2 Please describe a current or recent paid job whilst a student including hours a week worked and pay.
1.3 Have you done other paid jobs whilst a student at ____. Working as . . . ?
1.4 Did you do paid work before becoming a student? Working as . . . ?
1.5 Is anyone else in your household in paid work? Working as . . . ?
1.6 Ideally, what sort of paid work would you like to do after completing the course?
1.7 How do you feel your choices about and experience of paid work have affected/been affected by the fact you are a student? Give example/s.

2. **Housing**
2.1 What type of accommodation do you currently live in? (If you have just moved or are homeless, please discuss your most recent settled accommodation as a student.) Delete as appropriate: owner occupied/rented (local authority/housing association/private)/halls of residence/other (please describe).
2.2 Do/did you share this accommodation? Please indicate the numbers of adults and children you share/d with or put "halls." Number of adults: Number of children:
2.3 Please describe the layout of the accommodation (e.g. the nature and number of rooms). If in halls just describe your room.
2.4 What is/was the general condition of the accommodation? Delete as appropriate: Overcrowded/good condition throughout/mainly good but some defects/average-satisfactory/average but some deterioration/poor condition/very poor condition
2.5 Where do you/did you study in this accommodation? Please describe significant tensions and arrangements over quiet space to study.
2.6 What, if any, other accommodation have you lived in whilst studying at ____? Which was most conducive to study?
2.7 What accommodation did you live in before coming to ____?

3. **Income** (please give rough estimates)
3.1 What is the source of your income/household income? Indicate who in the household the income is paid to. If you don't know about an item and can't provide a rough estimate put "D/K." Paid work: Social security benefits: Student grant: Student loan: Access fund: Child support: Parental contribution: Other:
3.2 Has the amount of your/household income changed significantly since becoming a student? Please describe.
3.3 What access to and control of household income do you have?

4. **Domestic and household**
4.1 Who lives in your current household? Self only: delete: Y/N; Number of other adults and relationship to you: Number and ages of children and relationship to you: Is there anyone from the household absent e.g. overseas?
4.2 What housework commitments do you currently have? Are these shared commitments? Who with? Have these commitments changed significantly since you became a student?
4.3 What child-care commitments do you have? Are these shared commitments? Who with? Have these commitments changed significantly since you became a student?
4.4 Do you or anyone closely connected with you require personal assistance as a result of impairment or long-term sickness? Who gives this assistance and roughly what time does it involve?
4.5 Has short-term sickness—your own or another person's—had an impact on your studies? Please describe.
4.6 In general how has studying affected household and domestic commitments? Please describe any tensions arising from your need to study and significant arrangements which you made in order to study.

5. **Community, leisure and social activities**
5.1 What community, leisure and social activities are you currently involved in? Roughly what time do these involve?
5.2 Which of these would you identify as truly time for yourself, for example does other people's leisure impose upon your time?
5.3 Is this leisure involvement significantly different to that before you became a student? How?
5.4 Personal relationships and friendships
5.5 In general, in what significant ways do you feel studying has affected/been affected by your friendships and relationships? e.g. support from friends/relations e.g. friends/relations not understanding your need to study.

6. **Social position**
6.1 Please write a short pen picture about what social identities you see as relevant to yourself e.g. male/female, class, ethnicity, sexual orientation, colour etc.
6.2 By what educational route did you come into higher education (for example, school, access course etc.)?
6.3 Why have you chosen to come into higher education?

7. **Additional information**
7.1 Please add any additional information about other regular commitments on your time not identified above.
7.2 Please add any additional comments you would like to make.

Appendix 1b: Women Students' Reflective Log

Following from the longer questionnaire I gave out last year, this short questionnaire asks you to focus on the piece of work you have just handed in.

We are exploring some of the difficulties women students might encounter dealing with other commitments and tensions/conflicts whilst trying to study and produce work for college. We are particularly interested in finding out where and when women students actually study!

Don't worry that your answers may affect your mark for the essay . . . I promise we won't even look at the filled-in questionnaires until essays are marked. I was also a student who wrote some essays the night before they were due in!

As before, all responses will be treated with confidence and if I draw on the questionnaires I will change any identifying detail.

Dot Moss.

Name: _____ Course and date of entry: _____ D.o.b. ____

1. Think about the essay you have just handed in. Where did you study/write e.g. kitchen, living room, study, halls? Some or all of these?
2. Was this "your" space or did you have to move around?
3. Were there other activities taking place at the same time?
4. What time of the day did you study/write?
5. Was this a set time a day or did it vary?
6. What other commitments did you have? Domestic   Relationships   Social   Paid work
7. Can you identify specific tensions/conflicts over your study/writing in these areas Domestic   Relationships   Social   Paid work
8. If so how did you deal with them? What strategies did you adopt in order to complete the essay?   Domestic e.g. get help   Relationships e.g. row   Social e.g. stay in   Paid work e.g. go off "sick"
9. Is there anything you feel would have helped avoid any of these difficulties?
10. Please add any additional comments you would like to make about the process of writing the essay, in particular what space and times you used to write it.

Appendix 1c: Interview Schedule

Following the questionnaires you filled in before you completed the course I would like to interview some of you. During the interview I will ask you to reflect on your time at _____. I will be exploring your varied commitments (paid work, household and domestic, social and leisure and relationships), the strategies you used in order to study and your feelings and thoughts about coming into higher education. As before, all responses will be treated with confidence and if I draw on the interviews I will change any identifying detail.

Dot Moss.

1. **Background to entry**
1.1 What factors led you to come on the course? Probe: educational background/significant life events/hopes.
1.2 Were people close to you supportive of your plans? Probe: family/friends, support/resistance.
1.3 What significant practical arrangements were involved?   Housing: Domestic and household: Paid work: Child care and personal assistance:

2. **Paid employment**   Link and update questionnaires A & B.
2.1 Why did you choose this paid employment? Probe: space/distance/time/social background/ethnicity/other commitments/links curriculum.
2.2 Give an example of what the work actually involves. Describe a usual shift. Probe: time/space issues.
2.3 How have you handled tensions arising from studies and other commitments? E.g. essay deadlines, revision, child-care? Give example/s.
2.4 How did you feel about time spent in paid work in relation to your other commitments? How did others close to you feel? Give example/s.
2.5 Has paid work benefited your studies in any ways? E.g. have you studied at work/used resources from work etc.?

3. **Housing**   Link and update questionnaires A & B.
3.1 Focus on the same or other settled accommodation whilst at _____. Please sketch the layout of the accommodation, or room if in halls. If different accommodation to questionnaire A, ask appropriate questions about conditions/co-residents etc.

3.2 Using sketch plan, describe use of space and movements in the accommodation (or in halls) on a usual day (self and household).
3.3 Reflect on times when you needed to study/write and discuss/describe your uses of space at that time. What negotiations over quiet/safe space to study were significant?
3.4 How did you feel about studying at home/in halls? Probe: different times, when alone, when others about. Probe metaphor.

4. **Household and Domestic** Link and update questionnaires A & B.
4.1 What did these household and domestic activities involve? Jot down a typical day's activities. Probe time.
4.2 How have you handled tensions arising from studies? Who helped with housework/child care/personal assistance? Give examples.
4.3 How did you feel about household and domestic work? In relation to your need to study? Reflect on feelings at particular significant good/bad times.
4.4 In what ways if any has household and domestic work benefited your studies?

5. **Community, Leisure and Social** Link and update questionnaires A & B.
5.1 What things do you like/are you able to do to enjoy yourself? What do these involve? List. Probe time/space/place/feelings. (Your sex life doesn't have to come into this!)
5.2 When you need to study how are these activities affected?
5.3 How do you feel about time spent on these activities in relation to studies and other commitments?
5.4 Have these activities benefited your studies in any way?
5.5 Personal relationships. This is likely to overlap with other sections. May not be needed. Which people (relationship to you) gave you most practical support? Probe: time/space gains. Which people (relationship to you) gave you most emotional support? Probe: time/space gains. Which people (relationship to you) gave you least practical support? Probe: time/space losses. Which people (relationship to you) gave you least emotional support? Probe: time/space losses e.g. rows/being upset. How did you handle tensions when you needed to study?
5.6 In what ways do you feel you have changed during the course and in what ways have your personal relationships changed?

6. **The college and its expectations.** What are your general feelings about your college experiences? Give out sheet and discuss.

Course Requirements
Attendance
Essay deadlines
Examinations
Holidays
Probe: time (rigidity/flexibility)

Curriculum
course content
teaching styles
facilitation of your learning
connections to/distance from your experiences

Pastoral/Support/Advice
from tutors
from peers
from student services
EOPs and personal harassment issues
Probe: time gains/losses

Buildings/Spaces
large teaching rooms
small teaching rooms
computer room
common room
library
canteen
smoke room
Student's Union/bar/support groups
Grounds
Probe: uses of space/perceptions and feelings (e.g. comfort and safety) about space

7. **Advice to other women students** about time/space strategies.

Appendix 2
Research Data

Appendix 2a: Age Profile

The age profile of the sample, based on forty-four responses to a question about age, was as follows:

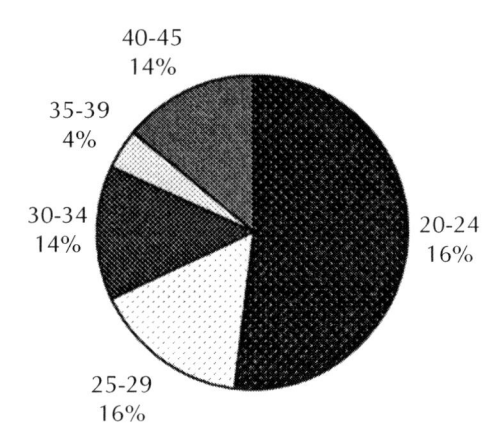

Appendix 2b: Previous Posts: Women Students

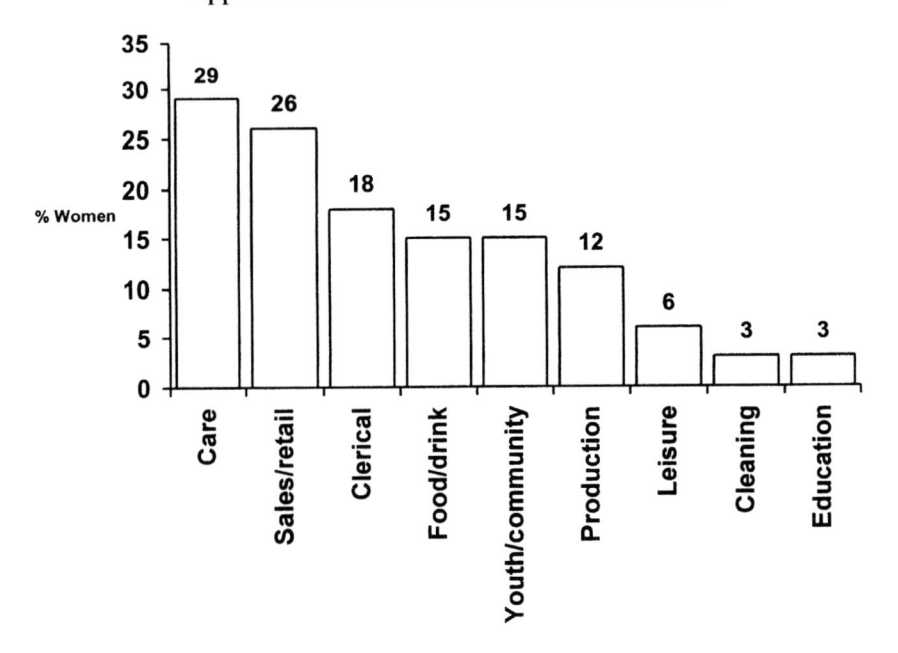

Category	% Women
Care	29
Sales/retail	26
Clerical	18
Food/drink	15
Youth/community	15
Production	12
Leisure	6
Cleaning	3
Education	3

Appendix 2c: Posts whilst Student

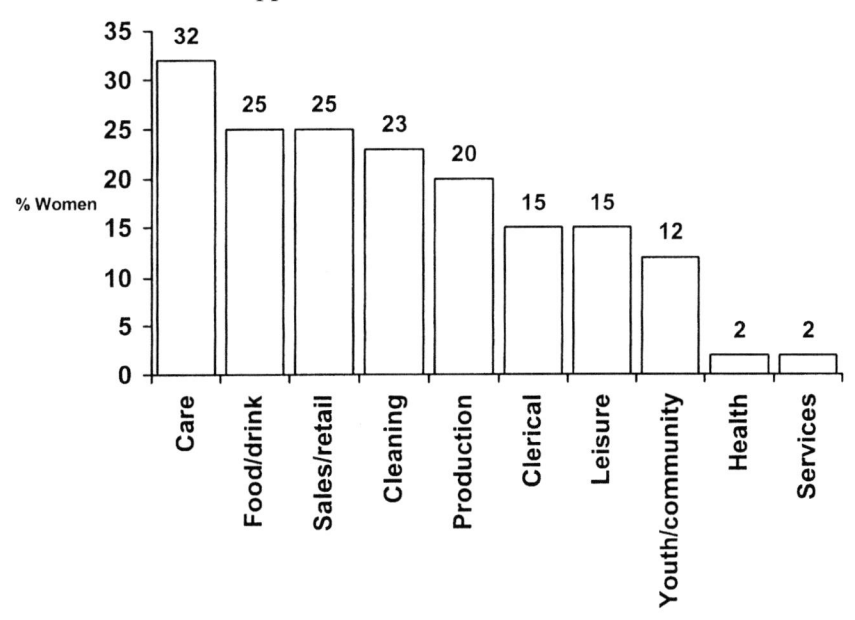

Category	% Women
Care	32
Food/drink	25
Sales/retail	25
Cleaning	23
Production	20
Clerical	15
Leisure	15
Youth/community	12
Health	2
Services	2

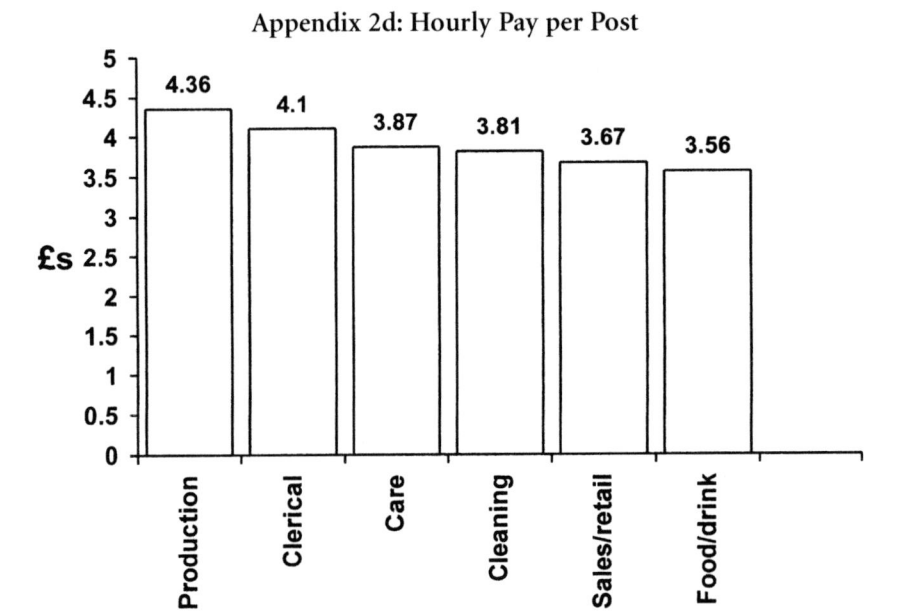

Appendix 2d: Hourly Pay per Post

Post	£s
Production	4.36
Clerical	4.1
Care	3.87
Cleaning	3.81
Sales/retail	3.67
Food/drink	3.56

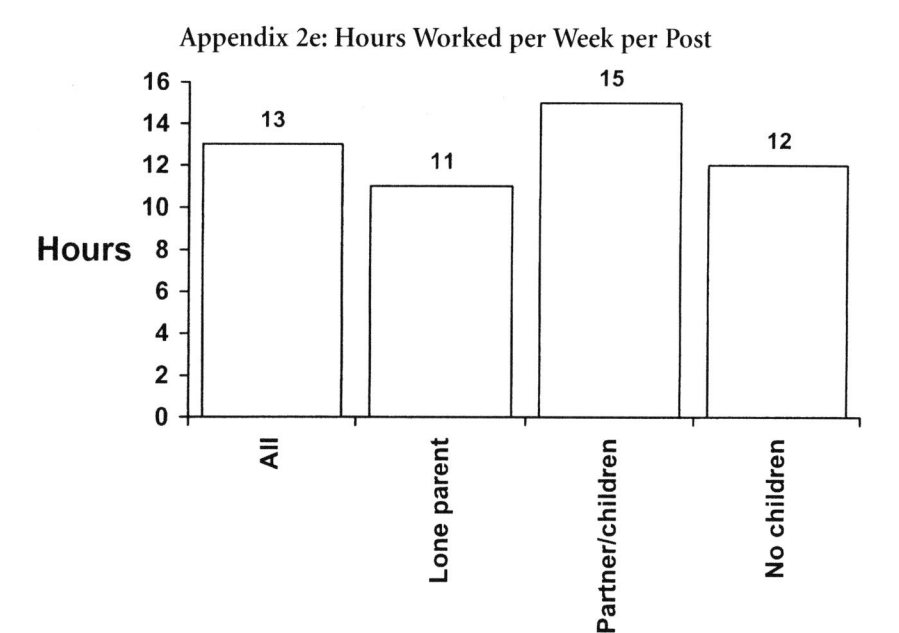

Appendix 2e: Hours Worked per Week per Post

Category	Hours
All	13
Lone parent	11
Partner/children	15
No children	12

Appendix 2f: Desired Posts

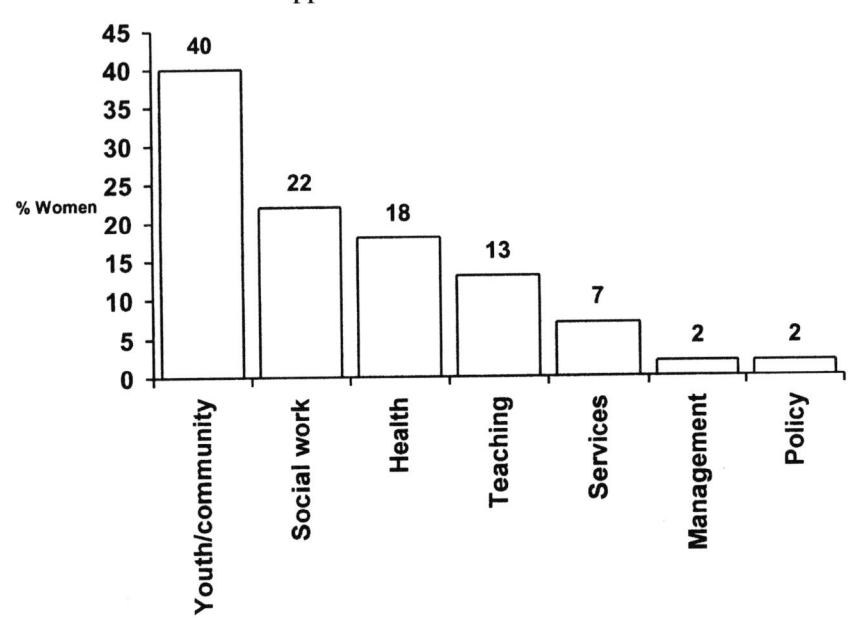

Post	% Women
Youth/community	40
Social work	22
Health	18
Teaching	13
Services	7
Management	2
Policy	2

Appendix 2g: Housing Status

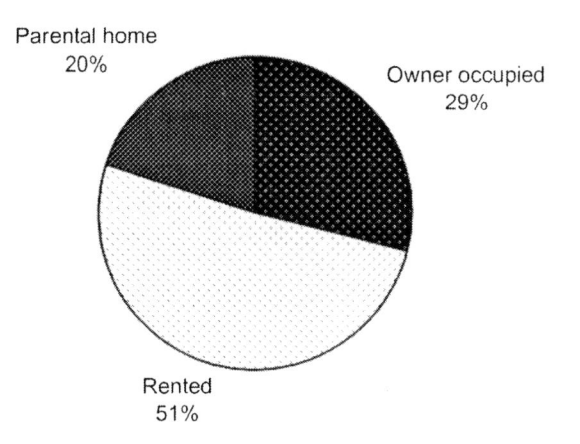

Parental home 20%

Owner occupied 29%

Rented 51%

Appendix 2h: Rented Status

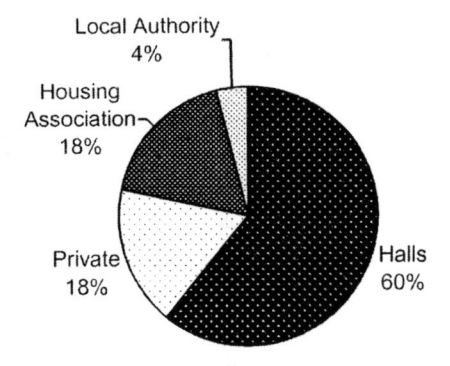

Appendix 2i: Housing Conditions: Women Students

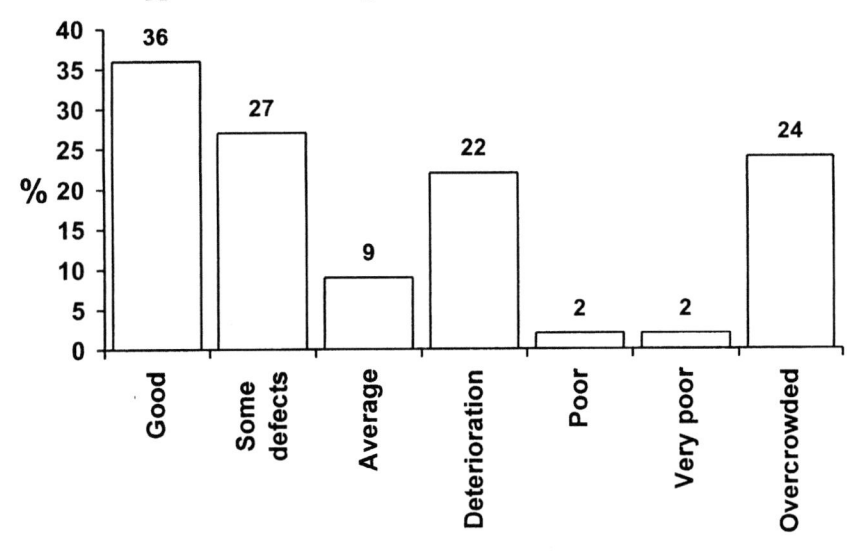

Condition	%
Good	36
Some defects	27
Average	9
Deterioration	22
Poor	2
Very poor	2
Overcrowded	24

Appendix 2j: Women Students' Households

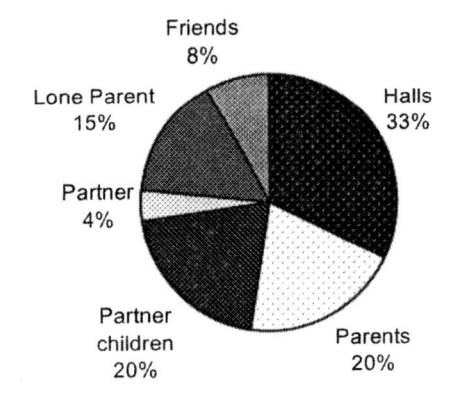

Appendix 2k: Women Students' Leisure

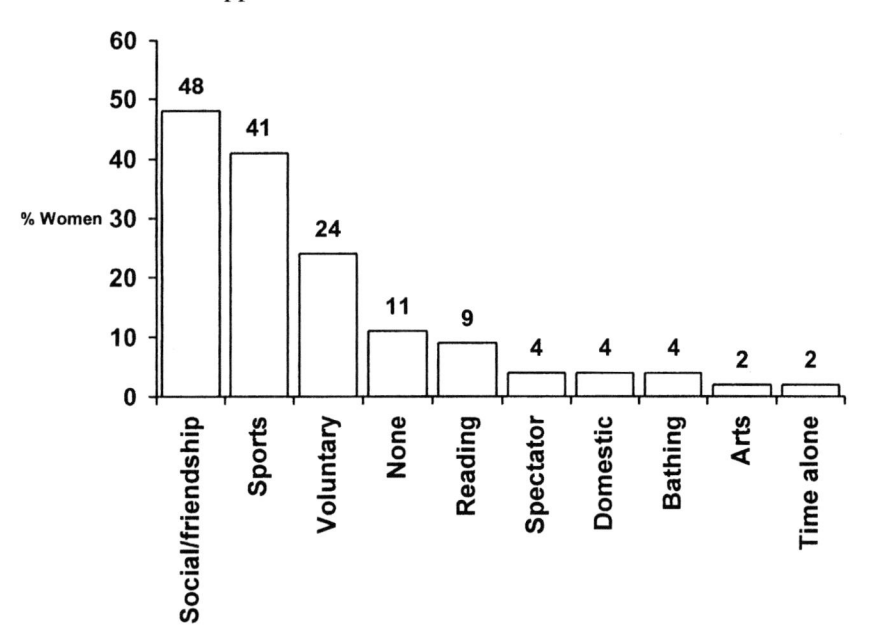

Activity	% Women
Social/friendship	48
Sports	41
Voluntary	24
None	11
Reading	9
Spectator	4
Domestic	4
Bathing	4
Arts	2
Time alone	2

Appendix 3
Research Proposal Extracts

Programme of Work

February 1998 to July 1998	Initial literature review. The literature review will continue until writing up. Formulate research questions. Research statement. P.G.C.R.M. R1.
July 1998 to December 1998	Write methodology working paper. Develop research materials.
January 1999 to April 1999	Use semi-structured questionnaire. Write key concepts working paper.
April 1999 to May 1999	Use reflective journal.
June 1999 to September 1999	Conduct semi-structured interviews.
September 1999 to December 1999	Organise and analyse data.
January 2000 to December 2000	Write transfer to PhD document.
January 2001 to January 2002	Theoretical development. Write remaining draft chapters.
January 2002 to July 2002	Final production of thesis.

Appendix 4

Brief Profiles of Women Students Who Were Interviewed

I have included in italics the ways in which women described themselves, followed by short cameos. Names and some other identifying details have been changed.

CLARE:
Female, white, British (mum Irish). No religion. Working class.
Clare was in her early twenties. She lived in halls of residence for most of the course but then moved out to live with her lesbian partner. She worked in a bar, a café and in a care home during the course, as well as being involved in community work.

DIANE:
Working class; "mongrel"; weird; physical health OK but still panic attacks.
Diane was in her early thirties and did considerable paid work during the course. She lived with her male partner in her own home, and gave assistance to both his and her mother, who both lived nearby. She had worked previously in management but had decided to change direction and do community based work.

GENI:
Female, heterosexual, white, no particular religious beliefs, size 10, blonde hair, caring, affectionate, student, sister, auntie, giggly.
Geni was in her early twenties. She came to college from her sister's home. She lived in halls of residence. She worked (during the college holidays) as a residential care worker, shop assistant and bar worker. Her main interest, both

academic and vocational, was holistic therapies. She had close friendships with women students in halls plus a boyfriend who was also at college.

HAZEL:
My ethnicity/identity is Afro-Caribbean. I was brought up in a working class family. Both parents worked in manual occupations.
Hazel did not say what her age was. She lived with her mother, who was very poorly during the course and at the time of interview was convalescing. Hazel's father died soon after the course started. All Hazel's siblings had left home. Her grandmother lived in a residential home and Hazel visited her regularly. She also took responsibility for a lot of household work. During the course Hazel did some library work but previously her experience was of very grim jobs in factories.

IRENE:
I am a white female aged twenty-two years and from a working class background. I am currently in a relationship of five years with my boyfriend.
Irene lived with her parents and sisters during the course. One of her sisters needed some personal assistance and Irene wanted to support her mother with this. Irene worked as a college receptionist, in a bar and as a care worker whilst studying. She said that she felt very close to her family and relied on her parents a lot when making decisions, however, she felt that her beliefs were changing due to her course of study and this distanced her from her family a little.

KATE:
Woman, white, Church of England/Scotland, "normal," working class, healthy, back trouble and epilepsy.
Kate was in her late forties. She had left school with no qualifications at fourteen and had previously worked in shops and factories. She was separated with three adult children, one of whom lived at home because of her severe impairments and need for constant support. Kate separated prior to coming onto the course. She owned her own house.

KELLY:
Female, African Caribbean.
Kelly was in her early thirties. She was a single parent with three young daughters. She had a strong reciprocally supportive family network of brothers and her mother, who needed some personal assistance and had mobility difficulties. Kelly had previous experience in community development but she did no paid work during the course. She lived in a property with a high rental and which was costly to heat.

LORNA:
Below average income; woman; no religion; white; good about self as a person.
Lorna was in her early thirties. She left school with few qualifications and said she had done very badly there. She worked as a sales assistant before returning to study. She had two children and separated from her male partner in the final year of the course.

MAURA:
Female, working class, white, British, heterosexual.
Maura was in her late twenties. She came down from Scotland with her male partner and three children to study. She did a considerable amount of paid work during the course as a support assistant and health care worker. She identified herself as the main decision-maker in the household. In the final year of the course she moved into halls of residence.

QASIR;
Female, black, non-religious, heterosexual, middle-aged, middle class origins, healthy, stress related illnesses.
Qasir was in her late thirties and lived with her husband and two children. She worked for many years in community development in Birmingham before the family moved north and she started the course. For a large part of the course she lived with her in-laws and gave personal assistance to them as well as doing casual jobs.

RACHEL:
Female, middle class, white, straight.
Rachel was in her early thirties. She had worked as an auxiliary in a psychiatric hospital for several years prior to the course. During the course she found work as an "out of hours" mental health worker in the community. She lived with her male partner who was a nurse. She had been assessed as dyslexic, but because of her paid work, found it difficult to access learning support.

RASHIDA:
Muslim, female, Pakistani, Asian, family—sister, daughter, friend.
Rashida was in her early twenties. She was the eldest of four children and lived in the parental home during the course. Her grandmother also lived there. Her younger brother had severe impairments and both he and grandmother needed personal assistance. As the eldest daughter, Rashida had a lot of domestic and household responsibilities, as well as providing personal assistance to these family members. She said she had not had the time to take up paid work during the course.

RIAN:

Female, mum and wife, middle class (values and beliefs—not income) white, heterosexual, sports person, enthusiastic.

Rian was in her late thirties. She lived with her male partner (who worked in computers) and three children. They owned their own house. Rian was from an agricultural background and worked as a labourer on her father's farm since childhood. She had a period of ill health and then got involved with sports activities. Her main interest was the health benefits of physical activity for women.

ROSE:

Female, student, friend, daughter.

Rose was in her early twenties. She came to college straight from school and lived in halls of residence. She did paid work during the holidays and had experience as a chambermaid and in retail. She had close women friends and a boyfriend, all of whom lived in halls. She had some health problems during the course.

SURAYA:

Asian, sister, colleague, student, auntie, woman, volunteer, daughter.

Suraya was in her early twenties when I interviewed her. She was the youngest child in her family and the only one still living in the parental home at that time. The family relied mainly on social security benefits because of her father's poor health. Suraya did a bit of paid work as a packer whilst a student. She also did considerable community work on an unpaid basis. She spent a lot of time whilst at college with Rashida.

SUSAN:

Female, wife, mother, student, white.

Susan was in her thirties. She left school at sixteen with very few qualifications. She was married to a joiner and had two school aged children. One of her sons had recently been identified as having learning disabilities and she spent a lot of time negotiating school support for him. Susan did a considerable amount of paid work whilst studying. She had formerly worked as a home care assistant and continued her evening work with a Social Services Department.

ZANDRA:

The fact that I'm [an] Afro Caribbean/woman, born in the West Indies and brought up in England. I realise the differences between black and white i.e. culture conflict and the fact that I'm still known as Afro Caribbean and my own experiences as a black working woman makes me realise that radical changes will take a very long time.

Zandra was in her early forties when I interviewed her. She was a lone parent with two daughters in their teens. Zandra had worked as a nurse for many years and during the course she maintained two posts, one in a small nursing home for older people, one in a larger home for people with Alzheimer's disease. She owned her own home and said that her main social activities were church and friends.

Bibliography

Acker, Joan, Kate Barry and Johanna Esseveld. "Objectivity and Truth: Problems in Doing Feminist Research." In *Beyond Methodology: Feminist Scholarship as Lived Research*, edited by Mary M. Fonow and Judith A. Cook. Bloomington: Indiana University Press, 1991.
Adam, Barbara. *Time and Social Theory.* Cambridge: Polity, 1990.
——— . *Timewatch, the Social Analysis of Time.* Cambridge: Polity Press, 1995.
——— . *Timescapes of Modernity: The Environment and Invisible Hazards.* London: Routledge, 1998.
Addelson, Kathryn Pyne. "The Man of Professional Wisdom." In *Beyond Methodology: Feminist Scholarship as Lived Research*, edited by Mary M. Fonow and Judith A. Cook. Bloomington: Indiana University Press, 1991.
Aitchison, Cara. "New Cultural Geographies: The Spatiality of Leisure, Gender and Sexuality." *Leisure Studies* 18 (1999): 19–39.
Alarcon, Norma. "The Theoretical Subject(s) of This Bridge Called My Back and Anglo-American Feminism." In *The Second Wave: A Reader in Feminist Theory*, edited by Linda Nicholson. London: Routledge, 1997.
Alcoff, Linda. "Cultural Feminism versus Post Structuralism: The Identity Crisis in Feminist Theory." In *The Second Wave: A Reader in Feminist Theory*, edited by Linda Nicholson. London: Routledge, 1988/1997.
Alldred, Pam. "Ethnography and Discourse Analysis: Dilemmas in Representing the Voices of Children." In *Feminist Dilemmas in Qualitative Research: Public Knowledge and Private Lives*, edited by Jane Ribbens and Rosalind Edwards. London: Sage, 1998.
Allen, John, and Doreen Massey. *Geographical Worlds.* Oxford: Oxford University Press, 1995.

Andermahr, Sonya, Terry Lovell and Carol Wolkowitz. *A Concise Glossary of Feminist Theory.* London: Arnold, 1997.

Anderson, Viv, and Jean Gardner. "Continuing Education in the Universities: The Old, the New and the Future." In *Educating Rita and Her Sisters: Women and Continuing Education,* edited by Roseanne Benn, Jane Elliot and Pat Whaley. Leicester: National Institute of Adult Continuing Education, 1998.

Ardener, Shirley, ed. *Women and Space: Ground Rules and Social Maps.* Oxford: Berg, 1993.

Arksey, Hilary, Ian Marchant and Cheryl Simmil. *Juggling for a Degree: Mature Students' Experience of University Life.* Unit for Innovation in Higher Education, Lancaster University, 1994.

Ashenden, Samantha. "Feminism, Postmodernism and the Sociology of Gender." In *Sociology after Postmodernism,* edited by David Owen. London: Sage, 1997.

Baden, Sally, and Anne Marie Goetz. "Who Needs (Sex) When You Can Have (Gender)? Conflicting Discourses on Gender at Beijing." *Feminist Review,* no. 5 (Summer 1997): 3–25.

Bagguley, Paul, Jane Mark-Lawson, Dan Shapiro, John Urry, Sylvia Walby and Alan Warde. *Restructuring: Place, Class and Gender.* London: Sage, 1990.

Bannerji, Himani. "But Who Speaks for Us? Experience and Agency in Conventional Feminist Paradigms." In *Unsettling Relations: The University as a Site of Feminist Struggles,* edited by Himani Bannerji, Linda Carty, Kari Dehli, Susan Heald and Kate McKenna. Toronto: Women's Press, 1991.

Bannerji, Himani, Linda Carty, Kari Dehli, Susan Heald and Kate McKenna. *Unsettling Relations: The University as a Site of Feminist Struggles.* Toronto: Women's Press, 1991.

Barrett, Michele. "Capitalism and Women's Liberation." In *The Second Wave: A Reader in Feminist Theory,* edited by Linda Nicholson. London: Routledge, 1997.

Belenkey, Mary F., Blythe McVicker, Nancy R. Goldburger, and J. M. Tarule. *Women's Ways of Knowing: The Development of Self, Voice and Mind.* New York: Basic Books, 1986.

Bell, Diane, and Renate Klein. *Radically Speaking: Feminism Reclaimed.* London: Zed Books, 1996.

Bell, Linda. "Public and Private Meanings in Diaries: Researching Family and Childcare." In *Feminist Dilemmas in Qualitative Research: Public Knowledge and Private Lives,* edited by Jane Ribbens and Rosalind Edwards. London: Sage, 1998.

Benn, Roseanne, Jane Elliot and Pat Whaley, eds. *Educating Rita and Her Sisters: Women and Continuing Education.* Leicester: National Institute of Adult Continuing Education, 1998.

Bennett, Fran, and Marilyn Howard. *Child Support: Issues for the Future.* London: Child Poverty Action Group, 1997.

Bhopal, Kalwant. *Gender, Race and Patriarchy: A Study of South Asian Women.* Aldershot: Ashgate, 1997.

Bignell, Kelly Coate. "Building Feminist Praxis Out of Feminist Pedagogy: The Importance of Students' Perspectives." *Women's Studies International Forum* 19, no. 3 (1996): 315–25.

Bird, John. *Black Students and Higher Education: Rhetorics and Realities.* Milton Keynes: Open University Press, 1996.

Borooah, Romy, Kathleen Cloud, Subodra Seshadri, T. S. Saraswathi, Jean T. Peterson and Amita Verma, eds. *Capturing Complexity: An Interdisciplinary Look at Women, Households and Development.* London: Sage, 1994.

Bottero, Wendy. "Clinging to the Wreckage? Gender and the Legacy of Class." *Sociology* 32, no. 3 (August 1998): 469–90.

Bourdieu, Pierre, Jean-Claude Chambordon and Jean-Claude Passeron. *The Craft of Sociology: Epistemological Preliminaries.* New York: Walter de Gruyter, 1968/1991.

Bowlby Sophie, Susan Gregory and Linda McKie. "Doing Home. Patriarchy, Caring and Space." *Women's Studies International Forum* 20, no. 3 (1997): 343–50.

Boyne, Roy. "Structuralism." In *The Blackwell Companion to Social Theory*, edited by Bryan S. Turner. Oxford: Blackwell, 1996.

Brewer, Rose. "Theorising Race, Class and Gender: The New Scholarship of Black Feminist Intellectuals and Black Women's Labour." In *Theorising Black Feminisms: The Visionary Pragmatism of Black Women*, edited by Stanlie M. James and Abena P. A. Busia. London: Routledge, 1993.

Bridge, Gary. "Mapping the Terrain of Time-Space Compression: Power Networks in Everyday Life." *Environment and Planning* D: Society and Space 15 (1997): 611–26.

Brown, Elsa B. "What Has Happened Here? The Politics of Difference in Women's History and Feminist Politics." In *The Second Wave: A Reader in Feminist Theory*, edited by Linda Nicholson. London: Routledge, 1997.

Bryant, Christopher, and David Jary, eds. *Giddens' Theory of Structuration: A Critical Appreciation.* London: Routledge, 1991.

Burden, Josephine. "Leisure, Change and Social Capital. Making the Personal Political." Paper presented at Leisure Studies Association International Conference "The Big Ghetto," July 1998.

Butler, Judith. "Imitation and Gender Insubordination." In *The Second Wave: A Reader in Feminist Theory*, edited by Linda Nicholson. London: Routledge, 1997.

Butler, Ruth, and Sophie Bowlby. "Bodies and Spaces: An Exploration of Disabled Peoples' Experiences of Public Space." *Environment and Planning* D: Society and Space 15 (1997): 411–33.

Callaway, Helen. "Spatial Domains and Women's Mobility in Yorubaland Nigeria." In *Women and Space: Ground Rules and Social Maps*, edited by Shirley Ardener. Oxford: Berg, 1993.

Callender, Claire. "Women and Employment." In *Women and Social Policy: An Introduction*, edited by Christine Hallett. London: Prentice Hall, Harvester Wheatsheaf, 1996.

Cannan, Crescy. "Enterprise Culture, Professional Socialisation and Social Work Education in Britain." *Critical Social Policy* 42 (Winter 1994/1995): 5–18.

Caplan, Paula J. *Lifting a Ton of Feathers: A Woman's Guide for Surviving in the Academic World.* University of Toronto Press, 1993.

Chouinard, Vera. "Making Space for Disabling Differences: Challenging Ableist Geographies." *Environment and Planning* D: Society and Space 15 (1997): 379–90.

Clegg, Sue. "The Feminist Challenge to Socialist History." *Women's History Review* 6, no. 2 (1997): 201–14.

Clegg, Sue. "Research Methodology Teaching Materials." Unpublished. Leeds Metropolitan University, 1998.

Clements, Luke. *Community Care and the Law*. London: Legal Action Group, 1997.

Cockburn, Cynthia. "The Material of Male Power." In *Waged Work: A Reader*, edited by Feminist Review Collective. London: Virago, 1986.

Coffey, Amanda, and Paul Atkinson. *Making Sense of Qualitative Data: Complementary Research Strategies*. London: Sage, 1996.

Cohen, Ira J. "Theories of Action and Praxis." In *The Blackwell Companion to Social Theory*, edited by Bryan S. Turner. Oxford: Blackwell, 1996.

Cooper, Davina. "Regard between Strangers: Diversity, Equality and the Reconstruction of Public Space." *Critical Social Policy* 18, no. 4 (November 1998): 465–92.

Connolly, Paul. "Racism and Post Modernism: Towards a Theory of Practice." In *Sociology after Postmodernism*, edited by David Owen. London: Sage, 1997.

Corrigan, Paul. 'The Trouble with Being Unemployed Is That You Never Get a Day Off." In *Freedom and Constraint: The Paradoxes of Leisure—Ten Years of the L.S.A.*, edited by Fred Coalter. London: Routledge, 1982/1989.

Corville, Cindy. "Re-examining Patriarchy as a Mode of Production. The Case of Zimbabwe." In *Theorising Black Feminisms: The Visionary Pragmatism of Black Women*, edited by Stanlie M. James and Abena P. A. Busia. New York: Routledge, 1993.

Cousins, Christine. Controlling Social Welfare: A Sociology of State Welfare Work and Organisation. Sussex: Wheatsheaf Books, 1987.

Crossley, Nick. "Emotion and Communicative Action: Habermas, Linguistic Philosophy and Existentialism." In *Emotions in Social Life: Critical Themes and Contemporary Issues*, edited by Gillian Bendelow and Simon J. Williams. London: Routledge, 1998.

Crouch, David, and Jan te Kloeze. "Camping and Caravanning and the Place of Cultural Meaning in Leisure Practices." In *Leisure, Time and Space: Meaning and Values in People's Lives*, edited by Sheila Scraton. Eastbourne: Leisure Studies Publications, 1998.

Davies, Karen. *Women and Time: The Weaving of the Strands of Everyday Life*. Aldershot: Avebury, 1990.

De-Groot, Joanne, and Mary Maynard, eds. *Women's Studies in the 1990s: Doing Things Differently*. Hampshire: Macmillan, 1993.

Deem, Rosemary. "Feminism and Leisure Studies: Opening Up New Directions." In *Relative Freedoms: Women and Leisure*, edited by Erica Wimbush and Margaret Talbot. Milton Keynes: Open University Press, 1988.

Deem, Rosemary. "Feminism and Leisure Studies." In *Sociology of Leisure: A Reader*, edited by Charles Critcher, Pete Bramham and Alan Tomlinson. London: E. and F. N. Spon, 1995.

Dehli, Kari. "Leaving the Comfort of Home: Working through Feminisms." In *Unsettling Relations: The University as a Site of Feminist Struggles*, edited by Himani Bannerji, Linda Carty, Kari Dehli, Susan Heald and Kate McKenna. Toronto: Women's Press, 1991.

Dominelli, Lena, and Ankie Hoogvelt. "Globalisation and the Technocratization of Social Work." *Critical Social Policy* 16, no. 2 (May 1996): 45–62.

Doucet, Andre. "Encouraging Voices. Towards More Creative Methods for Collecting Data on Gender and Household Labour." In *Gender Relations in Public and Private: New Research Perspectives*, edited by Lydia Morris and E. Stina Lyon. London: Macmillan, 1996.

Dragadze, Tamara. "The Sexual Division of Domestic Space among Two Soviet Minorities: The Georgians and the Tadjiks." In *Women and Space: Ground Rules and Social Maps*, edited by Shirley Ardener. Oxford: Berg, 1993.

Edwards, Rosalind. *Mature Women Students: Separating or Connecting Family and Education.* London: Taylor and Francis, 1993.

Edwards, Rosalind. "Access and Assets: the Experience of Mature Mother Students in Higher Education." In *Women and Social Policy: A Reader*, edited by Clare Ungerson and Mary Kember. London: Macmillan, 1997.

Edwards, Rosalind, and Simon Duncan. "Supporting the Family: Lone Mothers, Paid Work and the Underclass Debate." *Critical Social Policy* 53, no. 17 (November 1997): 29–49.

Elliot, Anthony. "Psychoanalysis and Social Theory." In *The Blackwell Companion to Social Theory*, edited by Bryan Turner. Oxford: Blackwell, 1996.

Elliot, Jane. "Locating Women: Theorising the Curriculum." In *Educating Rita and Her Sisters: Women and Continuing Education*, edited by Roseanne Benn, Jane Elliot and Pat Whaley. Leicester: National Institute of Adult Continuing Education, 1998.

Firestone, Shulamith. "The Dialectic of Sex." In *The Second Wave: A Reader in Feminist Theory*, edited by Linda Nicholson. London: Routledge, 1970/1997.

Fonow, Mary M., and Judith A. Cook. "Back to the Future." In *Beyond Methodology: Feminist Scholarship as Lived Research*, edited by Mary M. Fonow and Judith A. Cook. Bloomington: Indiana University Press, 1991.

Foster-Carter, Aiden. "The Sociology of Development." In *Sociology: New Directions*, edited by Mike Haralambos. Ormskirk: Causeway, 1985.

Foucault, Michel. *Madness and Civilisation: A History of Insanity in the Age of Reason.* London: Tavistock, 1989.

———. *Discipline and Punish: The Birth of the Prison.* Trans. Alan Sheridan. London: Penguin, 1977.

———. "Truth and Power." In *Power/Knowledge: Selected Interviews and Other Writings, 1972–1977.* Ed. Colin Gordon. New York: Pantheon Books, 1980.

———. *The History of Sexuality, Vol. 1.* Trans. Robert Hurley. New York: Vintage, 1980.

———. "The Eye of Power." In *Power/Knowledge: Selected Interviews and Other Writings, 1972–1977.* Ed. Colin Gordon. New York: Pantheon Books, 1980.

Frankenburg, Ruth. *White Women, Race Matters: The Social Construction of Whiteness.* London: Routledge, 1993.

Fraser, Nancy. "Structuralism or Pragmatics? On Discourse Theory and Feminist Politics." In *The Second Wave: A Reader in Feminist Theory*, edited by Linda Nicholson. London: Routledge, 1992/1997.

Gamarnikow, Eva, David Morgan, June Purvis and Daphne Taylorson, eds. *The Public and the Private*. England: Gower, 1983.

Giddens, Anthony. "Structuration Theory: Past, Present and Future." In *Giddens' Theory of Structuration: A Critical Appreciation*, edited by Christopher Bryant and David Jary. London: Routledge, 1991.

Gilroy, Paul. *Small Acts: Thoughts on the Politics of Black Culture*. London, Serpent's Tail, 1993.

Gilroy, Rose, and Roberta Woods. *Housing Women*. London: Routledge, 1994.

Glyptis, Sue. *Leisure and Unemployment*. Milton Keynes: Open University Press, 1989.

Goffman, Erving. *Asylums: Essays on the Social Situation of Mental Patients and Other Inmates*. London: Penguin, 1961.

Goode, Jackie, and Barbara Bagilhole. "The Social Construction of Gendered Equal Opportunities in UK Universities: A Case Study of Women Technicians." *Critical Social Policy* 18, no. 2 (May 1998): 175–92.

Gordon, Colin, ed. *Power/Knowledge: Selected Interviews and Other Writings, 1972–1977 by Michel Foucault*. New York: Pantheon Books, 1980.

Gorelick, Sherry. "Contradictions of Feminist Methodology." In *Race, Class and Gender: Common Bonds, Different Voices*, edited by Esther Ngan-Ling Chow, E. Doris Wilkinson and Maxine B. Zinn. London: Sage, 1996.

Green, Eileen. "Women Doing Friendship: An Analysis of Women's Leisure as a Site of Identity Construction, Empowerment and Resistance." Leisure *Studies* 17 (1998): 171–85.

Green, Eileen, Sandra Hebron and Diana Woodward. "Women, Leisure and Social Control." In *Women, Violence and Social Control*, edited by Jalna Hanmer and Mary Maynard. London: Macmillan, 1987.

Green, Eileen, Sandra Hebron and Diana Woodward. *Women's Leisure, What Leisure?* London: Macmillan, 1990.

Green, Eileen, and Sandra Hebron. "Leisure and Male Partners." In *Relative Freedoms: Women and Leisure*, edited by Erica Wimbush and Margaret Talbot. Milton Keynes: Open University Press, 1988.

Green, Eileen, and Diane Woodward. "Women's Leisure Today." In *Sociology of Leisure: A Reader*, edited by Charles Critcher, Pete Bramham and Alan Tomlinson. London: E. and F. N. Spon, 1995.

Griffiths, Vivienne. "From 'Playing Out' to 'Dossing Out': Young Women and Leisure." In *Relative Freedoms: Women and Leisure*, edited by Erica Wimbush and Margaret Talbot. Milton Keynes: Open University Press, 1988.

Gromet, Madeleine. "Conception, Contradiction and Curriculum." In *The Education Feminism Reader. Part III: Knowledge, Curriculum and Institutional Arrangements*, edited by Lynda Stone. London: Routledge, 1994.

Gurney, Craig M. "'Half of Me Was Satisfied': Making Sense of Home through Episodic Ethnographies." *Women's Studies International Forum* 20, no. 3 (1997): 373–86.

Gurvitch, Georges. "The Problem of Time." In *The Sociology of Time*, edited by John Hassard. London: Macmillan, 1964/1990.

Habermas, Jurgen. *The Theory of Communicative Action. Volume Two—Lifeworld and System: A Critique of Functionalist Reason*. Cambridge: Polity, 1981/1987.

Harding, Sandra. *The Science Question in Feminism*. Milton Keynes: Open University Press, 1996.

Hargreaves, Jennifer. *Sporting Females: Critical Issues in the History and Sociology of Women's Sports*. London: Routledge, 1994.

Hartmann, Heidi. "The Unhappy Marriage of Marxism and Feminism: Towards a More Progressive Union." In *The Second Wave: A Reader in Feminist Theory*, edited by Linda Nicholson. London: Routledge, 1981/1997.

Hartsock, Nancy. "The Feminist Standpoint: Developing the Ground for a Specifically Feminist Historical Materialism." In *The Second Wave: A Reader in Feminist Theory*, edited by Linda Nicholson. London: Routledge, 1983/1997.

Hassard, John, ed. *The Sociology of Time*. London: Macmillan, 1990.

Hemmings, Clare. "Bi-sexual Women's Spaces: In Between the Other Side or Round the Corner." Paper presented to Women's Studies Network Conference: Gendered Space, July 1998.

Henderson, Karla A., M. Deborah Bialeschki, Susan M. Shaw and Valerie J. Freyysinger. *Both Gains and Gaps: Feminist Perspectives on Women's Leisure*. State College, PA: Venture Publishing, 1996.

Hesse, Barnor. "Racism and Spacism in Britain." In *Tackling Racial Attacks*, edited by Peter Francis and Roger Matthews. Leicester: C. S. P. O. Publishers, 1993.

Heward, Chris. "Women and Careers in Higher Education: What Is the Problem?" In *Breaking Boundaries: Women in Higher Education*, edited by Louise Morley and Val Walsh. London: Taylor and Francis, 1996.

Hewitt, Patricia. *About Time: The Revolution in Work and Family Life*. Institute of Public Policy Research, London: Rivers Oram Press, 1993.

Hill Collins, Patricia. *Black Feminist Thought: Knowledge, Consciousness and the Politics of Empowerment*. London: Unwin Hyman, 1990.

Hirschon, Renee. "Essential Objects and the Sacred: Interior and Exterior Space in an Urban Greek Locality." In *Women and Space: Ground Rules and Social Maps*, edited by Shirley Ardener. Oxford: Berg, 1993.

Hochschild, Arlie. "The Sociology of Emotion as a Way of Seeing." In *Emotions in Social Life: Critical Themes and Contemporary Issues*, edited by Gillian Bendelow and Simon J. Williams. London: Routledge, 1998.

——— . "The Emotional Geography of Work and Family Life." In *Gender Relations in Public and Private: New Research Perspectives*, edited by Lydia Morris and E. Stina Lyon. London: Macmillan, 1996.

Hoff, Joan. "Gender as a Post Modern Category of Paralysis." *Women's History Review*, no. 2 (1994): 149–68.

Holliday, Ruth, Gayle Letherby, Lesli Mann, Karen Ramsey and Gillian Reynolds. "A Room of Our Own." In *Making Connections: Women's Studies. Women's Movements. Women's Lives*, edited by Mary Kennedy, Cathy Lubelska and Val Walsh. London: Taylor and Francis, 1993.

hooks, bell. *Talking Back: Thinking Feminist—Thinking Black.* London: Sheba, 1989.
———. *Feminist Theory: From Margins to Centre.* Boston, MA: South End Press, 1984.
———. *Killing Rage—Ending Racism.* New York: Henry Holt, 1996.
Jackson, Sue. "Safe Spaces: Women's Choices and Constraints in the Gendered University." Paper presented to Women's Studies Network Conference: Gendered Space, July 1998.
Jaggar, Allison M. "Love and Knowledge: Emotion in Feminist Epistemology." In *Women, Knowledge and Reality: Explorations in Feminist Philosophy*, edited by Ann Garry and Marilyn Pearsall. London: Routledge, 1996.
James, Cyril L. R. "Black Studies and the Contemporary Student." In *At the Rendezvous of Victory*, edited by Cyril L. R. James. London: Allison and Busby, 1969/1984.
James, Stanlie M., and Abena P. A. Busia, eds. *Theorising Black Feminisms: The Visionary Pragmatism of Black Women.* New York: Routledge, 1993.
Jaques, Elliot. "The Enigma of Time." In *The Sociology of Time*, edited by John Hassard. London: Macmillan, 1982/1990.
Jayaratne, Toby Epstein, and Abigail V. Stewart. "Quantitative and Qualitative Methods in the Social Sciences: Current Feminist Issues and Practical Strategies." In *Beyond Methodology: Feminist Scholarship as Lived Research*, edited by Mary M. Fonow and Judith A. Cook. Bloomington: Indiana University Press, 1991.
Kelly, Liz, Sheila Burton and Linda Regan. "Researching Women's Lives or Studying Women's Oppression? Reflections on What Constitutes Feminist Research." In *Researching Women's Lives from a Feminist Perspective*, edited by Mary Maynard and June Purvis. London: Taylor and Francis, 1994.
Koss, Mary P., and Hobart H. Cleveland. "Stepping on Toes: Social Roots of Date Rape Lead to Intractability and Politicisation." In *Researching Sexual Violence against Women: Methodological and Personal Perspectives*, edited by Martin D. Schwartz. Thousand Oaks, CA: Sage, 1997.
Layder, Derek. *New Strategies in Social Research.* Cambridge: Polity, 1993.
Lefebvre, Henri. *The Production of Space.* Oxford: Blackwell, 1991.
Letherby, Gayle, and Jen Marchbank. "To Boldly Go: Safe Spaces and Gendered Places in Feminist Research." Paper presented to Women's Studies Network Conference: Gendered Space, July 1998.
Lovell, Terry. "Feminist Social Theory." In *The Blackwell Companion to Social Theory*, edited by Bryan Turner. Oxford: Blackwell, 1996.
Lyon, M. L. "The Limitations of Cultural Constructionism in the Study of Emotion." In *Emotions in Social Life: Critical Themes and Contemporary Issues*, edited by Gillian Bendelow and Simon J. Williams. London: Routledge, 1998.
Madigan, Ruth, and Moira Monroe. "Gender, House, and 'Home': Social Meanings and Domestic Architecture in Britain." In *Women and Social Policy: A Reader*, edited by Clare Ungerson and Mary Kember. London: Macmillan, 1997.
Maguire, Meg. "In the Prime of Their Lives? Older Women in Higher Education." In *Breaking Boundaries: Women in Higher Education*, edited by Louise Morley and Val Walsh. London: Taylor and Francis, 1996.

Marshall, Barbara. *Engendering Modernity: Feminism, Social Theory and Social Change.* Cambridge: Polity, 1994.
Mason, Jennifer. "No Peace for the Wicked: Older Married Women and Leisure." In *Relative Freedoms: Women and Leisure,* edited by Erica Wimbush and Margaret Talbot. Milton Keynes: Open University Press, 1988.
Massey, Doreen. *Space, Place and Gender.* Cambridge: Polity Press, 1994.
Matrix. *Making Space: Women and the Man-made Environment.* London: Pluto Press, 1984.
Mauthner, Melanie. "Bringing Silent Voices into a Public Discourse: Researching Accounts of Sister Relationships." In *Feminist Dilemmas in Qualitative Research: Public Knowledge and Private Lives,* edited by Jane Ribbens and Rosalind Edwards. London: Sage, 1998.
Mauthner, Natasha, and Andrea Doucet. "Reflections on a Voice-Centred Relational Method. Analysing Maternal and Domestic Voices." In *Feminist Dilemmas in Qualitative Research: Public Knowledge and Private Lives,* edited by Jane Ribbens and Rosalind Edwards. London: Sage, 1998.
Maynard, Mary, and June Purvis. *Researching Women's Lives from a Feminist Perspective.* London: Taylor and Francis, 1994.
McDowell, Linda. *Capital Culture. Gender at Work in the City.* Oxford: Blackwell, 1997.
McGivney, Veronica. "Dancing into the Future: Developments in Adult Education." In *Educating Rita and Her Sisters: Women and Continuing Education,* edited by Roseanne Benn, Jane Elliot and Pat Whaley. Leicester: National Institute of Adult Continuing Education, 1998.
McKenna, Kate. "Subjects of Discourse: Learning the Language that Counts." In *Unsettling Relations: The University as a Site of Feminist Struggles,* edited by Himani Bannerji, Linda Carty, Kari Dehli, Susan Heald and Kate McKenna. Toronto: Women's Press, 1991.
Miller, Tina. "Shifting Layers of Professional, Lay and Personal Narrative. Longitudinal Childbirth Research." In *Feminist Dilemmas in Qualitative Research: Public Knowledge and Private Lives,* edited by Jane Ribbens and Rosalind Edwards. London: Sage, 1998.
Mirza, Heidi S. "Black Women in Education: A Collective Movement for Social Change." In *Black British Feminism: A Reader,* edited by Heidi Saffia Mirza. London: Routledge, 1997.
Morley, Louise, and Val Walsh, eds. *Breaking Boundaries: Women in Higher Education.* London: Taylor and Francis, 1996.
Morris, Jenny. *Pride against Prejudice: Transforming Attitudes to Disability.* London: Women's Press, 1991.
Moss, Dorothy, and Ingrid Richter. *Women Students: Creating Space and Time for Study.* Paper in Community Studies no. 16, Centre for Research in Applied Community Studies. Bradford, U.K.: Bradford College, 2000.
Moulton, Janice. "A Paradigm of Philosophy: The Adversary Method." In *Women, Knowledge and Reality: Explorations in Feminist Philosophy,* edited by Ann Garry and Marilyn Pearsall. London: Routledge, 1996.

Mowl, Graham, and John Towner. "Women, Gender, Leisure and Place: Towards a More 'Humanistic' Geography of Women's Leisure." *Leisure Studies* 14, no. 2 (April 1995): 102–16.

Naples, Nancy. *Grassroots Warriors: Activist Mothering, Community Work and the War on Poverty.* New York: Routledge, 1998.

Narayan, Uma. "Contesting Cultures: 'Westernisation,' Respect for Cultures and Third World Feminists." In *The Second Wave: A Reader in Feminist Theory*, edited by Linda Nicholson. London: Routledge, 1970/1997.

Nickols, Sharon, and Kamala Srinivasan. "Women and Household Production." In *Capturing Complexity: An Interdisciplinary Look at Women, Households and Development*, edited by Romy Borooah, Kathleen Cloud, Subodra Seshadri, T. S. Saraswathi, Jean T. Peterson and Amita Verma. London: Sage, 1994.

Oerton, Sarah. "Queer Housewives? Some Problems in Theorising the Division of Domestic Labour in Lesbian and Gay Households." *Women's Studies International Forum* 20, no. 3 (1997): 421–30.

Oliver, Mike, and Colin Barnes. *Disabled People and Social Policy: From Exclusion to Inclusion.* London: Longman, 1998.

Pahl, Jan. *Money and Marriage.* Basingstoke: Macmillan Education, 1989.

Palfreman-Kay, James M. "Disabled People and Access Opportunities into Higher Education." Paper presented at Higher Education Close Up: International Conference. Preston: University of Central Lancashire, 6–8 July 1998.

Parker, Stanley. "Towards a Theory of Leisure and Work." In *Sociology of Leisure: A Reader*, edited by Charles Critcher, Pete Bramham and Alan Tomlinson. London: E. and F. N. Spon, 1995.

Parr, Joan. "Theoretical Voices and Women's Own Voices: The Stories of Mature Women Students." In *Feminist Dilemmas in Qualitative Research: Public Knowledge and Private Lives*, edited by Jane Ribbens and Rosalind Edwards. London: Sage, 1998.

Pascall, Gillian. *Social Policy: A Feminist Analysis.* London: Tavistock, 1986.

Pascall, Gillian, and Roger Cox. "Education and Domesticity." *Gender and Education* 5, no. 1 (1993): 17–35.

Pyne Addelson, Katheryn. "The Man of Professional Wisdom." In *Beyond Methodology: Feminist Scholarship as Lived Research*, edited by Mary M. Fonow and Judith A. Cook. Bloomington: Indiana University Press, 1991.

Radice, Hugo. "From Warwick University Ltd. to British Universities Plc." *Red Pepper* (March 2001).

Ragin, Charles C. *Constructing Social Research: The Unity and Diversity of Method.* Thousand Oaks, CA: Pine Oaks Press, 1994.

Reinharz, Shulamit, with Lynn Davidman. *Feminist Methods in Social Research.* New York: Oxford University Press, 1992.

Report of the Commission on Social Justice. *Social Justice: Strategies for National Renewal.* London: Vintage, 1994.

Ribbens, Jane, and Rosalind Edwards, eds. *Feminist Dilemmas in Qualitative Research: Public Knowledge and Private Lives.* London: Sage, 1998.

Ribbens, Jane, and Rosalind Edwards. "Living on the Edges: Public Knowledge, Private Lives and Personal Experience." In *Feminist Dilemmas in Qualitative Research: Public Knowledge and Private Lives*, edited by Jane Ribbens and Rosalind Edwards. London: Sage, 1998.

Roberts, Marion. *Living in a Man-made World: Gender Assumptions in Modern Housing Design*. London: Routledge, 1991.

Rocke, Cynthia, Susanne Torre and Gwendolyn Wright. "The Appropriation of the House: Changes in House Design and Concepts of Domesticity." In *New Space for Women*, edited by Gerda R. Wekerle, Rebecca Peterson and David Morley. Boulder, CO: Westview Press, 1980.

Rose, Gillian. *Feminism and Geography: The Limits of Geographical Knowledge*. Cambridge: Polity Press, 1993.

Rubin, Gayle. "The Traffic in Women: Notes on the 'Political Economy' of Sex." In *The Second Wave: A Reader in Feminist Theory*, edited by Linda Nicholson. London: Routledge, 1975/1997.

Saegart, Susan, and Gary Winkel. "The Home: A Critical Problem for Changing Sex Roles." In *New Space for Women*, edited by Gerda R. Wekerle, Rebecca Peterson and David Morley. Boulder, CO: Westview Press, 1980.

Sarwasthi, T. S. "Women in Poverty Contexts." In *Capturing Complexity: An Interdisciplinary Look at Women, Households and Development*, edited by Romy Borooah, Kathleen Cloud, Subodra Seshadri, T. S. Saraswathi, Jean T. Peterson and Amita Verma. London: Sage, 1994.

Sayer, Andrew. *Method in Social Science: A Realist Approach*. London: Routledge, 1992.

Scraton, Sheila, and Beccy Watson. "Gendered Cities: Women and Public Leisure in the 'Post Modern City.'" *Leisure Studies* 17, no. 2 (April 1998): 123–37.

Scraton, Sheila, Pete Bramham and Beccy Watson. "Going Out: Elderly Women, Leisure and the Postmodern City." In *Leisure, Time and Space: Meaning and Values in People's Lives*, edited by Sheila Scraton. Eastbourne, U.K.: Leisure Studies Publications, 1998.

Sennett, Richard. *Flesh and Stone: The Body and the City in Western Civilisation*. London: Faber and Faber, 1994.

Shaw, Susan. "Conceptualising Resistance: Women's Leisure as Political Practice." In *Journal of Leisure Research* 33, no. 2 (2001): 186–201.

Shelton, Beth A. *Women, Men and Time: Gender Differences in Paid Work, Housework and Leisure*. New York: Greenwood Press, 1992.

Sibley, David. *Geographies of Exclusion*. London: Routledge, 1995.

Silva, Elizabeth de. *Good Enough Mothering? Feminist Perspectives on Lone Motherhood*. London: Routledge, 1996.

Silverman, David. *Interpreting Qualitative Data*. London: Sage, 1993.

Skar, Sarah L. "Andean Women and the Concept of Space/Time." In *Women and Space: Ground Rules and Social Maps*, edited by Shirley Ardener. Oxford: Berg, 1993.

Skeggs, Beverley. *Formations of Class and Gender: Becoming Respectable*. London: Sage, 1997.

Smart, Barry. "Post-modern Social Theory." In *The Blackwell Companion to Social Theory*, edited by Bryan Turner. Oxford: Blackwell, 1996.

Smith, Dorothy. *The Everyday World as Problematic: A Feminist Sociology*. Milton Keynes: Open University Press, 1987.

Sorokin, Pitrim, and Robert Merton. "Social-Time: A Methodological and Functional Analysis." In *The Sociology of Time*, edited by John Hassard. London: Macmillan, 1937/1990.

Spain, Daphne. *Gendered Spaces*. Chapel Hill: The University of North Carolina Press, 1992.

Stanley, Liz. "Feminist Praxis and the Academic Mode of Production." In *Feminist Praxis*, edited by Liz Stanley. London: Routledge, 1990.

Stanley, Liz, and Sue Wise. *Breaking Out Again: Feminist Ontology and Epistemology*. London: Routledge, 1993.

———. "Method, Methodology and Epistemology in Feminist Research Processes." In *Feminist Praxis*, edited by Liz Stanley. London: Routledge, 1990.

———. "Feminist Research, Feminist Consciousness and Experiences of Sexism." In *Beyond Methodology: Feminist Scholarship as Lived Research*, edited by Mary M. Fonow and Judith A. Cook. Bloomington: Indiana University Press, 1991.

Stanley, Liz. "Historical Sources for Studying Work and Leisure in Women's Lives." In *Relative Freedoms: Women and Leisure*, edited by Erica Wimbush and Margaret Talbot. Milton Keynes: Open University Press, 1988.

Tait, Ann. "The Mastectomy Experience." In *Feminist Praxis*, edited by Liz Stanley. London: Routledge, 1990.

Thomas, Kim. *Gender and Subject in Higher Education*. Buckingham: The Society for Research into Higher Education and Open University Press, 1990.

Thrift, Nigel. *Spatial Formations*. London: Sage, 1996.

Thrift, Nigel, and Peter Williams, eds. *Class and Space: The Making of Urban Society*. London: Routledge and Keegan Paul, 1987.

Tongue, J. "New Packaging, Old Deal? New Labour and Employment Policy Innovation." *Critical Social Policy* (May 1999): 217–32.

Urry, John. "Time and Space in Giddens' Social Theory." In *Giddens' Theory of Structuration: A Critical Appreciation*, edited by Christopher Bryant and David Jary. London: Routledge, 1991.

———. "Sociology of Time and Space." In *The Blackwell Companion to Social Theory*, edited by Bryan S. Turner. Oxford: Blackwell, 1996.

Van Every, Jo. "Understanding Gendered Inequality: Reconceptualising Housework." *Women's Studies International Forum* 20, no. 3 (1997): 411–20.

Wallace, Claire. "Between the State and the Family: Young People in Transition." *Youth and Policy* 25 (1988): 25–36.

Walsh, Val. "Feminism Matters." *Woman Bulletin* 43. London: Association of University Teachers, 1998.

Watts, Carol. "Time and the Working Mother: Kristeva's 'Women's Time' Revisited." *Radical Philosophy*, no. 91 (September/October 1998): 6–17.

Wearing, Betsy. *Leisure and Feminist Theory*. London: Sage, 1998.
Wekerle, Gerda R., Rebecca Peterson and David Morley. *New Space for Women*. Boulder, CO: Westview Press, 1980.
Whaley, Pat. "Women Researching in Continuing Education: Voices, Visibility and Visions." In *Educating Rita and Her Sisters: Women and Continuing Education*, edited by Roseanne Benn, Jane Elliot and Pat Whaley. Leicester: National Institute of Adult Continuing Education, 1998.
Whannel, Garry. "Electronic Manipulation of Time and Space in Television Sport." In *Leisure, Time and Space: Meaning and Values in People's Lives*, edited by Sheila Scraton. Eastbourne, U.K.: Leisure Studies Publications, 1998.
White, Ann. "Perception of Student Needs: A Feminist Approach." Paper presented at Higher Education Close Up conference. University of Central Lancashire, 6–8 July 1998.
White, Jacquelyn W., and John A. Humphrey, "A Longitudinal Approach to the Study of Sexual Assault. Theoretical and Methodological Considerations." In *Researching Sexual Violence against Women: Methodological and Personal Perspectives*, edited by Martin D. Schwartz. London: Sage, 1997.
Williams, Fiona. "Somewhere Over the Rainbow: Universality and Diversity in Social Policy." *Social Policy Review* 4. Ed. Nick Manning and Robert Page. Canterbury: Social Policy Association (1992): 200–219.
Williams, Peter. "Constituting Class and Gender: A Social History of the Home, 1700–1901." In *Class and Space: The Making of Urban Society*, edited by Nigel Thrift and Peter Williams. London: Routledge and Keegan Paul, 1987.
Wimbush, Erica. "Mothers Meeting." In *Relative Freedoms: Women and Leisure*, edited by Erica Wimbush and Margaret Talbot. Milton Keynes: Open University Press, 1988.
Wright, Susan. "Place and Face: Of Women in Doshman Ziari, Iran." In *Women and Space: Ground Rules and Social Maps*, edited by Shirley Ardener. Oxford: Berg, 1993.
Young, Iris. "Beyond the Unhappy Marriage: A Critique of Dual Systems Theory." In *Women and Revolution: The Unhappy Marriage of Marxism and Feminism. A Debate on Class and Patriarchy*, edited by Lydia Sargent. London: Pluto Press, 1981.
Yuval Davis, Nira. *Gender and Nation*. London: Sage, 1997.
Zmroczek, Christine, and Pat Mahoney. "Women's Studies and Working-Class Women." In *Desperately Seeking Sisterhood: Still Challenging and Building*, edited by Magdalene Ang-Lygate, Chris Corrin and Henry Millsom. London: Taylor and Francis, 1997.

Index

Main sections or substantial references are indicated by **bold type**. Tables are indicated by *italics*.

access courses, 176
Acker, Joan, 81, 93
action: concepts in social theory, 28; emotion and action, **28–29**; power and action, 28, **29–30**. *See also* centres of action
Adam, Barbara, 4 *bis*, 5, 6, 31–32, 35–36, 37 *bis*, 39, 41, 49, 63, 88, 92, 171, 202
adversarial method, 10, 12, 82
afro-centric feminist epistemology, 10
age, 35, 66, **101–2**, 151, 163–64, 189, 198, 200, *229*
Aitchison, Cara, 39
Alcoff, Linda, 18–19
Anderson, Viv, 203
archaeology of knowledge, 15, 20
Ardener, Shirley, 34, 37–38, 86–87, 88
Aries, Philippe, 52
Ashenden, Samantha, 16
assessment of establishments, 169, 196, 207
Assiter, Alison, 21–22

Bannerji, Himani, 80
Barrett, Michele, 13
Barry, Kate, 81, 93
Beijing conference (1995), 17
Bell, Linda, 91
benefits. *See* welfare
Bhopal, Kalwant, 69
black women. *See* race
body: image, 147; theorisation of, 34–35; work concerned with, 131
Bottero, Wendy, 20–21, 22
Bourdieu, Pierre, 14, 18, 34, 37, 39–40, 80
Bowlby, Sophie, 34–35, 137
Brewer, Rose, 14
Burden, Josephine, 175
"bureaucratisation" (Habermas), 29
Busia, Abena, 50
Butler, Judith, 16
Butler, Ruth, 34–35

Callender, Claire, 109
Cannan, Crescy, 63

— 263 —

capitalism: concepts of, 21, 29; influence on higher education, 63–64, 72, 169; involves manipulation of space, 37; and paid work, 128; and patriarchy, 12, 13; and value of time, 30; and women's oppression, 11–13
care work, **128–30, 134–35**, 151, 194. *See also* childcare
Carer's Allowance, 104
Carter, Aiden Foster, 62
case studies: Barbara, 104; Christine, 114, 117, 177; Clare, 139, 161, 164 *bis*, 165, 177 *bis*, 179, 188, **243**; Diane, 137, 164, 165, 165–66, 177, 178–79, 181 *bis*, 183, 186, 188, 189, **243**; Geni, 113, 117, 118, 132, 134 *bis*, 137, 137–38, 146, 147, 158, 159, 161, 162, 182, 212, **243–44**; Grace, 106, 113; Hazel, 135, 136–37, 138, 139, 142, 145, 157, 172, 174–75, 175, 183, 189, **244**; Irene, 129, 134, 139, 140, 146, 161, 174, 175, 178, 213, **244**; Janice, 113; Kate, 148–49, 157, 159, 163–64, 180–81, 182, 186, 189, **244**; Katrina, 104, 110, 115, 117; Kelly, 102, 118, 140, 141, 143, 144, 161–62, 162, 163, 175, 176 *bis*, 177 *bis*, 180 *bis*, 181, 184 *bis*, 187, 189, **244**; Lorna, 140, 146, 147, 158, 164, 166–67, 182, 183, 189, **245**; Maggie, 105, 143; Martha, 143; Mary, 102, 137, 140–41, 158, 164, 174; Maura, 133, 148, 162 *ter*, 163 *bis*, 172, 173, 174, 176, 183, 184, 189, 212, **245**; Maureen, 107, 114, 174, 175; Michelle, 113; Olivia, 105, 112; Prabha, 106, 187; Qasir, 140, 142–43, 146–47, 156, 181, 182, 184, 187, 188, **245**; Rachel, 110, 129, 134, 159, 162, 163, 172–73, 173, 176, **245**; Rashida, 128, 135, 138, 139, 142, 145, 146, 157, 158, 162, 162–63, 174, 186, 186–87, 213, **245**; Rian, 104, 106–7, 131–32, 133, 134, 142, 146, 147, 148, 158 *bis*, 161, 162, 166, 167, 173, 174, 175 *bis*, 178, 180, 185, 188, **246**; Rose, 112, 130, 137–38, 147, 157, 159, 161, 161–62, 162 *bis*, 163, 184, 213, **246**; Rowena, 115; Ruth, 131; Sally, 118; Sonya, 113, 118; Suraya, 110, 128, 130, 131, 133, 134, 139, 140, 145, 157, 158, 162–63, 176, 185, 213, **246**; Susan, 107–8, 130, 134, 135, 138, 140, 142, 143, 145, 148 *bis*, 157, 159, 161, 173, 176, 179, 180, 184, 188, 189, 212, **246**; Zandra, 106, 129, 129–30, 137, 141–42, 143, 147, 148 *bis*, 157, 164, 173, 179, 184, 187, **246–47**
centres of action, women as, **53–56, 171–91**; concept of, 5, 6, **39–41**, 78; concept vs. *agency* and *praxis*, 206; creating space and time to study, 3, **70–71**; regaining ground and meaning, **201–2**
Child Support Agency, 104, 141, 195
childcare, **114**, 115, 139–40, 180, 181; for disabled child, 182, 186
childhood, concepts of, 139
Christmas, 156
"claims to truth" (Foucault), 14–15, 20
class, 11, 32, **102–3**, 151
Clegg, Sue, 18–19
Cockburn, Cynthia, 13
Cohen, Ira, 28, 205
Coit, Elizabeth, 54
Collins, Patricia Hill, 10, 49
"communicative action" (Habermas), 29
community, **116–17**, 150, **175**; significance for black women, 117, 119, 141, 143–44, 151, 175. *See also* friendship
Connolly, Paul, 16, 34
Cook, Judith, 87
Cooper, Davina, 51
Corrigan, Paul, 52, 117
Cousins, Christine, 109
Cox, Roger, 69
"critical modernism" (Marshall), 17
critical realism, feminist, 4, **17–22**
Crossley, Nick, 30
curriculum, 64, **66–67**, 69, 72, 164

Davies, Karen, 4, 30, 37, 38, 49–50, 54, 55, 69, 86, 86–87, 88, 92, 181, 198, 212
deadlines, 156, 157, 160, 164, 183–84, 202, 214–15
Deem, Rosemary, 50, 119
Dehli, Kari, 71–72
dependency theory, 62
Derrida, Jacques, 14, 15
"dialogue" (Yuval Davis), 21
disability: disability theory, 34–35; dyslexia, 159, 165, 176; hearing impairments, 159; and paid work, 133; problems for carers, 139, 182, 186; problems for disabled students, 70, 159, 160–61, 180–81; self-identity as disabled persons, **104–5**, 106, 151–52; unawareness of rights, 157–58. *See also* illness; mental health problems
"discourse" (Foucault), 14–15, 20; "subjugated discourses," 21
domestic violence. *See* violence against women
Dominelli, Lena, 64
Doshman Ziari (Iran), 54
Duncan, Simon, 29
dyslexia, 159, 165, 176

Edwards, Rosalind, 29, 69, 70–71, 93, 141, 199, 211
emotion: and action, **28–29**; author's (reflective log), 211, 216; emotional dissonance, 39, 53, 128, *136, 144, 149,* 150, *167,* 168, 188–89, **201**; fear of failure, 176; guilt, 143, 144, 148, 189; importance in research, 19, 30, 36, 87, 88, 204; reactions to studying/end of studies, 174, 185; transcribing of, 92. *See also* emotional landscapes
emotional landscapes: of higher education, **165–68**; of housing and household, **142–44**; of leisure, **148–49**; of paid work, **133–35**
employment. *See* work, paid

enterprise culture, 63, 72, 159
epistemologies, 10, 12, 13, **78–84, 208–9,** **213–14**
Esseveld, Johanna, 81, 98
ethical issues, 5, **93–94**
ethnicity. *See* race
ethnomethodology, school of, 28
Every, Jo Van, 50

femininity, ideas about, 9, 133
feminist theory, **9–10**; adversary method, 10, 12; critical realism, 17–22; material analyses of women's oppression, 10–14; need for new concepts, 20–22; need for "real world research," 18–19, 78; post-structural analyses and, 10, **14–16**, 17; reflexive approach, **19–20**, 79–80, 209; relation between theory and research, 10; research into women and higher education, 2–3; theorising from women's life experiences, 3
"field" (Bourdieu), 34
finances: of higher education institutions, 158, 159, 169, 196, **197–98**; of labour market, 194; of students' grants and loans, 65 *bis,* 103, 104, 108, 132, 195, 207; of students' home and household, 86, 141–42, 151, 195, 198; of students' wider income, **103–4**. *See also* welfare; work, paid
Firestone, Shulamith, 11
Fonow, Mary, 87
Foucault, Michel, 14–15, 18, 52
Frankenburg, Ruth, 16
Fraser, Nancy, 14, 15
friendship: and leisure, **118–19**, 146; and paid work, 135; significance of, 55, 121

Gardner, Jean, 203
gay people, 16, 177. *See also* lesbians
gender, concepts of, 17
Giddens, Anthony, 32–33, 34, 36, 37, 129, 204

Gilroy, Paul, 53, 84
Glyptis, Sue, 49
Gorelick, Sherry, 3, 19
grants and loans. *See* finances
Green, Eileen, 50, 55, 92, 175
Gromet, Madeleine, 71, 203
"grounded theory" approach, 85
"ground rules" (Ardener), 37
Gurvitch, Georges, 33, 204

Habermas, Jurgen, 29–30, 197
"habitus" (Bourdieu), 34, 37
Hagerstrand, Torsten, 32
halls of residence, 104–5, 111, 112, 113, 115, **115–16**, 121, 137, 137–38, 146, 151, 161, 174, 182, 195, 200–201, 213
Harding, Sandra, 18, 77
Hargreaves, Jennifer, 52–53, 147
Hartmann, Heidi, 12, 13
Hartsock, Nancy, 12–13
Hassard, John, 30, 31, 33
hearing impairments, 159
heating problems, 112, 113, 116, 187
Henderson, Karla, 117, 145
Hesse, Barnor, 49
Heward, Chris, 203
Hewitt, Patricia, 55, 85
higher education: ambiguous meaning, 197–201; author's experience as student (reflective log), **208**; collective nature of, **158**; concepts of, 81, 150, 199–200; critical questions and issues, 55–56, 178, 179, 185, 190; curriculum, 64, **66–67**, 69, 72, 164; deadlines, 156, 157, 160, 164, 183–84, 202, 214–15; emotional landscape of, **165–68**, *167*; feminist perspectives, 62–63; finances of, 158, 159, 169, 196, **197–98**; individuality, enterprise and production, 157, **159**; learning environment, **67**, **69–70**; merging of campuses, 159; pathways to and through, **68–69**, 171, **172–78**, **207–8**; physical landscape of, **160–63**; political perspectives of, 61–62, **63–65**, 168–69, **194–97**; regulation and reforms of, 61, 63, 64, 72, **159**, 168–69,195–96; rhythm of, 172, **179–85**, **202**; rigidity of temporal rationales, 156, **157–58**; semesterisation, 156, 157 *bis*, **158**, 159; social divisions and, **158–59**; social landscape of, **163–65**; source of leisure, 119 ; spatial/temporal practices of, **63–65**, **156–57**. *See also* teaching in higher education
Hirschon, Renee, 51
Hochschild, Arlie, 29, 39, 54, 188–89
home: author's experience (reflective log), **209–20**; concepts of, 121, 138, 146, 166–67; connections to studies, 178–79; men's/women's viewpoint, 51–52, 53; as private space, 49; relation to paid work, 54; separation from workplace, 30–31. *See also* housing and household
Hoogvelt, Ankie, 64
hooks, bell, 16, 69, 69–70, 143
housing and household: and capitalism, 11; conditions, **112**; domestic commitments and studies, 49–50, **114–16**, **195**; emotional landscape of, 142–44, *144*, 150; household time, 138; households, persons in *238*; housework, 11, **114**, 137, 138, 139, 142, 179; housing conditions, *237*; housing design, **51–52**, 54, 137, 201; housing market, 120–21; housing status, *235, 236*; pathways to higher education through, 173–75; physical landscape of, **136–39**; public housing, 195; salient features, **116**; social landscape of, **139–42**; and study space, **112–13**; summary of findings; tenure, **111–12**; transitions, 113. *See also* finances; home

identity: age and, **101–2**; class and, **102–3**; colour and geographical heritage and, **103**; identity politics,

21–22; illness and disability and, **105**; religion and, 105; salient features, **105**–7; sexuality and, **105**; social construction of, 101; wider income and, **103**–4

illness, **104**–5, 183, 210, 211, 215. *See also* disability

"imperatives for action" (Habermas), 29, 152

income. *See* finances

individual and the social, 34, 37

interactionist school, 28

Jackson, Sue, 67, 71, 164
Jacques, Elliot, 31, 37
Jagger, Allison, 28–29, 39, 212
James, C. L. R., 67
James, Stanlie, 50
Jayaratne, Toby Epstein, 84, 87

Kelly, Liz, 92

Lacan, Jacques, 11, 14, 15
"landscapes," 39, 40, 82–83, 91, 93, 127–28
lecture theatres, 67, 161–62
Lefebvre, Henri, 4–5, **33**–**34**, 36–37, 37, 38, **40**–**41**,120, 121, 149, 171, 181
leisure: activities, **116**–17, *239*; author's experience (reflective log), **210**–11; author's research set in leisure studies, 83; concepts of, 48, 52–53, 54, 121, 146, 147, 150, 199; emotional landscape of, **148**–**49**, *149*; feminist leisure theory, 55; free time may not be leisure, 117, 119; and friendships, 55, 117, **118**–**19**; and higher education, 55, 83; pathways to higher education through, *175*; physical landscape of, 51, **145**–46; political aspect, 55; relation to studies, 200; relation to work, 50, 52, 55; research method, 92; social landscape of, **146**–47; time for self/selfish time, 50, 83, 146, 147, 150, 200

lesbians: and childcare, 139; harassment/hostility towards, 164, 188; identity as, 16, 105; mutual support, 177

Lévi-Strauss, Claude, 11

library, 67, 156, 159, 162, 183

"life-world" (Habermas), 29

linguistic theory, 15

"*longue durée* of institutions" (Giddens), 32, 129

"lunacy" (Foucault), 15

Madigan, Ruth, 51
Marshall, Barbara, 14, 17, 19, 20
Marxist theory: and centres of action, 40; and feminist theory, 10, 11–14 *passim*, 15; Habermas and, 29; and segregation, 48; and space/time, 30, 31, 33–34, 37
Massey, Doreen, 3, 31, 35, 36
materialist theory, **10**–**14**, 17; relevance to research, **79**
McDowell, Linda, 51, 131
McKenna, Kate, 165
men: dominance of male viewpoint, 17, 67; male issues, 164; male partners, 70, 104, 109, 118, 138, 166–67, 173, 174, 175, 182, 188, 199, 200, 210; male violence, 140–41, 209–10; South Asian men's attitudes, 142–43. *See also* patriarchy
mental health problems, 105, 177
Merleau-Ponty, Maurice, 34
Merton, Robert, 31
methodology of research: inductive/deductive, 84–85; quantitive/qualitative, 6, 79, **84**–**88**, 91, 92, 100
methods of research, **88**–**94**
Mirza, Heidi, 68
modernisation theory, 62
Monroe, Moira, 51
mother students: attitudes of, 190; author's experience (reflective log), 210, 211; family support for, 118; and

learning environment, 70, 112; and leisure, 117, 121; lone parents, 115; mothering devalued, 69–70; and paid work, 29, 108, 109; quality of mothering, 139–40, 143, 144, 150, 188, 199; self-identification as mothers, 105. *See also* childcare
mothers of students, 139–40, 180
motivation for studying, **68–69**, 101, 165–66, 168, 178, 179, 184, 187
Moulton, Janice, 10
Mowl, Graham, 51
multi-functional spaces, 54, 137, 160, 173, 201

Narayan, Uma, 69
New Labour, 64, 194, 196, 197
New Right, 194
noise problems, 113, 115, 116, 161, 182

ontology, realist, 18
oppression and inequality of women: denial of, 17; feminist focus on, 2, 13, 78–79; materialist analyses of, **10–14**, 17, 20; resistance to, 55; search for foundations of, 20; systems of (*see* capitalism; patriarchy); traffic in women, 11. *See also* violence against women

paid work. *See* work, paid
Parker, Stanley, 52
Parr, Joan, 80, 87, 165
Parsons, 28
Pascall, Gillian, 69
patriarchy, **12–13, 20, 21**
"phallic drift" (Bell and Klein), 17
place: concept of, 33 *bis*, 186; place for research (author's reflective log), 216–27; place for studies, 172, **186–90**, 199, 211–13
"polyhedron" (Williams), 20, 22
"positionality" (Alcoff), 18
post-structuralism: concepts of, 20; and feminism, **14–16**, 17, **19–20**; impasse with structuralism, 21; and racialisation, 16; relevance to research, **79**
power: academic, 216–17; complexity of relationships, 14, 14–15; concepts of, 28, 36; in the household, 4, 54, 198; physical, 13; regimes and structures of, 16, **19, 20–22**, 78–79, **204**; social power and action, **29–30**; space and, 33
privacy, 49
productivity, 131, 159
psychoanalytical theory, 12, 15, 204–5
public sector restructuring, **194–95**
"punishment" (Foucault), 15

Quality Assessment Agency, 196

race: Afro-Caribbean heritage, 10, 106; black studies, 66–67; concepts of, 16, 21, 53, 84; and identity, **103**, 177; inequality and racism, 68–69, 144, 163; and nationalist positions, 53; and paid work, 14, 50; segregation, 131, 138, 162–63; self-identification, 103, 106; significance of community, 117, 119, 141, 143–44, 151; solidarity within ethnic groups, 177; South Asian heritage, 69, 111, 128; South Asian men's attitudes, 142–43; visibility of black women's lives, 49
Radice, Hugo, 168–69
"realism" (Sayer), 17
reason, theoretical emphasis on, 28, 30, 205
reflexivity, **19–20, 79–80**, 209
Reinharz, Shulamit, 10
relativism, 21
religion: Christian/Muslim calendars, 159, 186; extreme beliefs, 164; Judeo-Christian, 30; and leisure/community activity, 103, 117, 119; self-identification in terms of, 105
research, author's: analysis and findings, 6–7, 93, 100, **193–218**; background

and development, 1–7, 22; epistemology, 78–81, 83–84; ethical issues, 5, 93–94; interviews, 91–92, 127–52, 155–69, 212–13, 226–28; methodology, 6, 79, 84–88, 91, 92, 100; methods, 88–94; questionnaire, 90, 99–122, 221–24; questions, table of, 72; reflective log, author's, 83, 90, 206–17; reflective log, students', 91, 171, 225; sample, 6, 88–90, 100 (*see also* case studies); spatial/temporal practices of, 81–82; spatial/temporal representations of, 82–83; transcribing, 92–93; uniqueness of, 7

research, feminist. *See* feminist theory

research, further needed, 217–18; need for "real world research," 18–19, 78

Research Assessment Exercise (RAE), 169, 207

rhythm: concept of, 41, 93, 171, 180, 205; of higher education/studies, 172, **179–85, 202**; of research (author's reflective log), **214–16**; rhythmicity of life, 193; rhythmicity of organisms, 36; symphony of rhythms, 180, 203

Ribbens, Jane, 93

Roberts, Marion, 51

Rose, Gillian, 36

Rubin, Gayle, 11–12

Runa culture, 42

Sayer, Andrew, 17, 81

"schemes of perception" (Ardener), 34

school as pathway to higher education, 176

segregation/separation: gender-based, **48–49**, 48, 63, 66, 85, 86; of housing communities, 138; racial, 131, 138, 162–63; special schooling, 138

self-esteem, 189

semesterisation, 156, 157 *bis*, **158**, 159

sexuality: nineteenth-century definitions, 15; homosexuality, 16, 177 (*see also* lesbians); issues related to, 152, 164–65; pressure to be sexually attractive, 147; root of social/economic divisions, 11; self-identification in terms of, 16, **105**; sex life first thing lost under pressure, 181

Shaw, Susan, 55

Shelton, Beth Anne, 86

Skar, Sarah Lund, 42

Skeggs, Beverley, 14, 68, 209, 216–17

sleep, 211

Smith, Dorothy, 3, 14, 87

smokers, 177

"snapshots," 38, 88, 90, 93, 100, 155

social action theory, 36, 208–9

social care work, **128–30, 134–35**, 151, 194

social divisions, 11, 66, **79**, **158–59**, 164

social housing provision, 112

"social maps" (Ardener), 37–38

social position: and curriculum, 66, 164; and financial expectations, 198; and identity, **101–7**; impact on students, 70, 187; and leisure, 117, 119; shapes decisions, 68; spatial/temporal factors increase visibility, 79

social power. *See* power

social security benefits. *See* welfare

"social space" (Bottero and Williams), 21, 22

Sorokin, Pitkim, 31

space: concepts of, 3–4, 31, 37, 38; field, habitus and body, **34–35**, 37; gendered, 67; Lefebvre's theories (*see* Lefebvre); multi-functional spaces, 54, 137, 160, 173, 201; others' space, 50; patterns of spatial use, 172, **178–79**; place distinguished from, 33 *bis*; place for research (author's reflective log), **216–27**; place for studies, 172, **186–90**, 199, **211–13**; as process not object, 51; public/private, 49; and research methodology, **85–88**; women's creation of (*see*

centres of action). *See also* spatial/temporal
Spain, Daphne, 3–4, 48, 63, 85
spatial/temporal concepts, **3–4, 30–32,** 33, **35–36,** 36–7, *42*, 81, **204–6**
spatial/temporal factors: normative assumptions, **200–201**; in social research, 3–4; visibility of, 32, 36, 41, 217
spatial/temporal practices: of author's research, 81–82; concept of, 5, **37–38**; gender and, **48–50**; of higher education, **63–65, 156–60**; of housing/household, **111–16**; and leisure/community, **116–19**; of paid work, **107–11**; research questions, 50; and social position/identity, **101–7**;
spatial/temporal representations: of author's research, 82–83; concept of, 5, **38–39**; gender and, **51–53**; and higher education, **66–68, 160–68**; of housing and household, **136–44**; of leisure, **145–49**; of paid work, **128–36**; research questions, 53
staffing. *See* teaching in higher education
"standpoint," concept of, 21
Stanley, Liz, 10, 19, 20, 54, 77, 85
Stewart, Abigail V., 84, 87
stress, 148
structuralism, 10, 14, 16, 17, 21. *See also* post-structuralism
structuration theory, **32–33**
student, concept of, 81
Students' Loans Company, 195
student union building, 160, 162; bar, 116, 137–38, 146, 162–63
student union movement, 169
support for students, 118, 121, 140, 177, 207. *See also* friendship
surveillance, 194, 195
"system" (Habermas), 29, 30

"Taylorisation," 64
teaching in higher education: methods, 67; quality of, 165; role of academics, 207, 209; staff/student ratios, 157, 158, 160; staff workloads, 207; tutor/student relations, 65, 156, 157, 158, 176–77, 207–8, 212–13
Teaching Quality Assessment (TQA), 169
texts, 15, 20, 214
Thomas, Kim, 66
Thompson, E. P., 30, 197
Thrift, Nigel, 34 *bis*
time: financial value of, 115, 132–33; free time (*see* leisure); in higher education (*see* higher education); household time (*see* housing and household); "mapping" of, 33; multiple time spans, 5, 36, 40; others' time, 31, 37, **49–50,** 54; **181–82**; qualitative time, 31; and research methodology, **85–88**; simultaneous activities, 86; time for self (see leisure); time squeeze, 108; "timescape" (Adam), 39; women's creation of (*see* centres of action). *See also* rhythm; spatial/temporal; time, concepts of
time, concepts of: astronomical time, 31; clock time/clock-led routines, 30, 38, 65, 92, 128, 180, 202; complexity of, 38; dualist thinking, 35; hierarchy of values, **199–200**; linear time, 30, 35; male/female concepts, 86; multiplicity of meanings, 4, 31–32, 87; ontological time, 31; "planes of temporality" (Giddens), 32, 129; in social theory, **30–32**; "temporal rationality" (Davies), 38
Towner, John, 51
"transversalism," concept of, 21–22
tutor-student relations. *See* teaching in higher education

universalism, 21
Urry, John, 4, 30, 32, 33, 36

violence against women, 48, 53, 140, 141, 209–10

Watts, Carol, 205
Wearing, Betsy, 52, 145
Weber, Max, 28, 29, 48
welfare: benefits, 104 *bis*, 194–95; deregulation of, 194–95; restructuring of, 109; Welfare State, 64; workfare systems, 64
Whaley, Pat, 214
White, Ann, 62, 68, 69
Williams, Fiona, 20, 21, 22
Williams, Peter, 30–31, 49
Wimbush, Erica, 55
Wise, Sue, 20, 77, 85
women: concept of, 18, 19, 80; experiences of higher education, 68–71, 203–4; experiences of regimes of power, 204; experiences, theorising from, 3; ideas of femininity, 133, 147; women's agency (*see* centres of action). *See also* oppression and inequality of women; violence against women
Women's Studies, 66
Woodward, Diane, 92
work, concepts of: "real work" seen as physical labour, 128, 134; value of paid work vs. education/housework, 115, 132–33; work ethic, 64
work experience placements, 110, 179, 207
work, in the home: domestic commitments and studies, 114–16; housework, 11, **114**, 137, 138, 139, 142, 179
work, paid: author's work (reflective log), 114–16; author's landscapes of work and study (reflective log), 212–13; cleaning and domestic, 131; and dominant practices, 55; emotional landscape of, **133–35**, *136*; employment opportunities, 108, 120; future times (aims), **110–11**, *234*; gendered, 108–9; and the home, 54; hourly pay, *232*; hours worked, *233*; negative perceptions of, 110, *136*; as others' time, 50; past times (before studying), **107**, *230*; pathways to higher education through, **172–73**, 207–8; physical landscape of, **128–31**; place of friendship, 146; present times (while studying), **107–10**, *231*; restructuring of, 109, 194; retail sector, 130–31; shift system, 128; social care, **128–30**, **134–35**, 151, 194; social landscape of, **131–33**
work, voluntary, 55
workfare, 64
workplace, separation of home from, 30–31

Yerania (Athens), 51
Young, Iris, 13
Yuval Davis, Nira, 19–20, 21–22, 101, 206

About the Author

Dorothy Moss is currently Senior Lecturer in Childhood Studies at Leeds Metropolitan University. Her research is in the area of feminist sociology and the sociology of space and time. She completed her PhD at Leeds in 2002. She has a BA (Hons.) in English Language and Literature from Liverpool University (1972) and an MA in Social and Community Studies from Bradford University (1984).

She has worked in higher education since 1987 and has experience lecturing in the areas of Social Welfare Law, Social Policy, Sociology, Welfare Rights and Community Practice. Her work prior to becoming an academic was community-based. She worked in advocacy at a number of rights projects including Women's Aid, Citizen's Advice Bureaux and Tribunal Assistance Projects. She has also been involved in many community-based projects concerned with welfare rights on a voluntary basis.